Green Energy and Technology

Climate change, environmental impact and the limited natural resources urge scientific research and novel technical solutions. The monograph series Green Energy and Technology serves as a publishing platform for scientific and technological approaches to "green"—i.e. environmentally friendly and sustainable—technologies. While a focus lies on energy and power supply, it also covers "green" solutions in industrial engineering and engineering design. Green Energy and Technology addresses researchers, advanced students, technical consultants as well as decision makers in industries and politics. Hence, the level of presentation spans from instructional to highly technical.

Indexed in Scopus.

Indexed in Ei Compendex.

Iwona Bąk · Katarzyna Cheba
Editors

Green Energy

Meta-analysis of the Research Results

Editors
Iwona Bąk
Faculty of Economics
West Pomeranian University of Technology
in Szczecin
Szczecin, Poland

Katarzyna Cheba
Faculty of Economics
West Pomeranian University of Technology
in Szczecin
Szczecin, Poland

ISSN 1865-3529　　　　　　　　ISSN 1865-3537 (electronic)
Green Energy and Technology
ISBN 978-3-031-12530-0　　　　ISBN 978-3-031-12531-7 (eBook)
https://doi.org/10.1007/978-3-031-12531-7

© The Editor(s) (if applicable) and The Author(s), under exclusive license to Springer Nature Switzerland AG 2023
This work is subject to copyright. All rights are solely and exclusively licensed by the Publisher, whether the whole or part of the material is concerned, specifically the rights of translation, reprinting, reuse of illustrations, recitation, broadcasting, reproduction on microfilms or in any other physical way, and transmission or information storage and retrieval, electronic adaptation, computer software, or by similar or dissimilar methodology now known or hereafter developed.
The use of general descriptive names, registered names, trademarks, service marks, etc. in this publication does not imply, even in the absence of a specific statement, that such names are exempt from the relevant protective laws and regulations and therefore free for general use.
The publisher, the authors, and the editors are safe to assume that the advice and information in this book are believed to be true and accurate at the date of publication. Neither the publisher nor the authors or the editors give a warranty, expressed or implied, with respect to the material contained herein or for any errors or omissions that may have been made. The publisher remains neutral with regard to jurisdictional claims in published maps and institutional affiliations.

This Springer imprint is published by the registered company Springer Nature Switzerland AG
The registered company address is: Gewerbestrasse 11, 6330 Cham, Switzerland

Preface

Climate change constitutes a significant challenge—organisational, economic, scientific and technological—for society, the state and its administrative structures. The public debate on this subject is focused not just on the benefits but also the costs connected with implementing specific initiatives in this direction. This concerns the policy of the state, hinging on the dilemma between technological progress and economic growth, and environmental pollution, while public opinion calls for the effective management of natural resources, which means using them according to the principles of sustainable development.

Energy supply and the use of energy produced with means generating less CO_2, gas emissions and no toxic waste play the key role in achieving the objectives of climate protection. Hence, the term 'green energy' which refers to environment-friendly electricity produced from renewable sources of energy such as water, wind, biomass and the sun.

In recent years, energy policy has become an area of active political involvement on the part of many countries. Due to the activities of public institutions favouring green energy, there have been changes in the attitudes and behaviour of both private individuals and entire communities, as well as companies and financial institutions. Hence, there is a growing need for a widely understood discussion on the subject of the support (political, legal, financial and social) and development of green energy. These issues, among others, are the subject of the research whose results are presented in this book. The approach used by the authors is problem based in order to explain the complexity of the described phenomena, as well as to point to the areas important in further debate, which concern the international situation and the interests of all stakeholders in the energy market.

The added value of the research presented in the book is the multidimensional approach to the issue of green energy. Each chapter addresses the matter from diverse perspectives, including financial, technological and social. The results also consider various approaches to the collection and processing of data on green energy. The research used, among others, the technique of a systematic review based on earlier reports and studies treated as the source material. The authors also carried out analyses of data from official records and databases, as well as the analysis of primary data obtained directly for the needs of the conducted study. The research employed the methods of descriptive statistics (parameters of structure, correlation and dynamics) and advanced methods of multidimensional comparative analysis.

The first chapter presents a review of the most important definitions concerning green energy, its kinds and sources. It also addresses the main development trends, paying particular attention to the impact on the natural environment and energy, and energy effectiveness.

The second chapter presents the political issues related to green energy. The use of green energy in the economy is at present one of the leading subjects in the political debate conducted both at EU level and in individual member states of the European Union. The main objective, namely the inevitable drift towards climate neutrality, is fairly obvious; however, one should also draw attention to the implementation of this goal in legal regulations dedicated to individual sectors of the modern economy.

In the third chapter, the role of green energy in the world's economic development is discussed. Currently, there is observed a growing demand for energy accompanying economic growth. Bearing in mind the ever faster decline of natural resources, and at the same time the growing costs of their exploitation, it is obvious that there is an ongoing search for such sources of energy that respect the economic aspects of generating energy as well as being focused on the protection of the natural environment.

The fourth chapter presents the issues connected with the impact of green energy on the protection of the natural environment—both from the viewpoint of its benefits and the problems resulting from the common use of green technologies.

The fifth part discusses the social aspects of the green transformation. The chapter also addresses the social conditions of the green transformation, seen through the prism of both highly developed countries and economies just starting on the path to such a transformation.

The directions of financing the green energy transformation are the topic of the next chapter. This part of the book presents the most relevant possibilities of financing the green transformation, also providing examples of similar regulations introduced in EU member states.

Green energy transformation models are widely presented in the next part of the book. The pace of the green transformation is highly diversified depending on the level of the economic development of individual countries. The authors trayed to elaborated some heuristic models for group of countries with similar level of development in the area of green transformation.

Finally, the results of the study of disproportions in the area of green energy in EU countries are discussed in the last part of the book. The authors used selected methods of statistical multivariate analysis as the research tool, which allowed for the classification of the objects (EU countries) according to the level of green energy transformation and distinguish groups of countries at a similar level of development.

Szczecin, Poland
Iwona Bąk
Katarzyna Cheba

Contents

Green Energy—A Review of the Definitions and the Main
Directions of Development 1
Iwona Bąk and Katarzyna Cheba

Green Energy in the Political Debate 17
Agnieszka Malkowska and Arkadiusz Malkowski

The Role of Green Energy in the Economic Growth of the World 41
Irena Łącka

Green Energy and Its Impact on Environmental Protection 59
Beata Szczecińska

Selected Social Aspects of the Green Transformation 71
Anna Barwińska-Małajowicz and Miroslava Knapková

The Directions of Financing the Green Energy Transformation 85
Anna Spoz and Magdalena Ziolo

Green Energy Transformation Models—Main Areas and Further
Directions of Development 105
Katarzyna Cheba and Iwona Bąk

The Study of Disproportions in the Area of Green Energy in EU
Countries 123
Maciej Oesterreich and Katarzyna Wawrzyniak

Conclusion 151

Green Energy—A Review of the Definitions and the Main Directions of Development

Iwona Bąk and Katarzyna Cheba

Abstract The chapter presents a review of the most important definitions concerning green energy, its kinds and sources. It also addresses the main development trends, paying particular attention to the impact on the natural environment and energy, and energy effectiveness. Another vital issue discussed here is the transition from an economy based on traditional sources of energy, towards one based on sustainable sources of energy, the so-called green transformation of an economy. The chapter also addresses matters concerning global energy security. Moreover, attention is drawn to the growing tendency of investing in renewable energy in EU countries.

1 Introduction

In the course of the last two centuries, the human population has increased sevenfold. By 2011 it reached the figure of seven billion, and is expected to increase by two billion over the next 25 years (Van Bavel 2013). The increase has been accompanied by the development of science and technology, and unprecedented urbanisation. Energy has become a vital need of society (Kumar et al. 2020), with huge increase in demand, accompanied by the reduction of available natural resources, including fossil fuels. Thus diminishing resources, together with excessive demand for energy, have created the problem of energy security. The enormous growth in population numbers and demand for energy have affected the natural environment. Its deteriorating condition, increasingly noted around the 1950s, has directed the attention of the world's governments to issues linked with raising ecological awareness and counteracting environmental threats. The United Nations (UN) is considered to be the chief initiator of actions for the protection of the natural environment, which over the last few decades has undertaken a string of initiatives aimed at embedding

---

I. Bąk (✉) · K. Cheba
Faculty of Economics, West Pomeranian University of Technology in Szczecin, Szczecin, Poland
e-mail: iwona.bak@zut.edu.pl

K. Cheba
e-mail: katarzyna.cheba@zut.edu.pl

© The Author(s), under exclusive license to Springer Nature Switzerland AG 2023
I. Bąk and K. Cheba (eds.), *Green Energy*, Green Energy and Technology,
https://doi.org/10.1007/978-3-031-12531-7_1

the concept of sustainable development in the consciousness of the governments and populations of an ever-increasing number of countries. The term 'green economy' is inseparably linked with the paradigm of sustainable development based on technologies playing a role subordinated to the natural environment and corporate social responsibility for the quality of life of future generations. A green economy is focused on the perception of the threats resulting from the expansive economic and social activity of humankind, irreversibly damaging both the natural environment and its limited resources (Söderholm 2020).

The functioning and progress of the economy are mostly based on the use of fossil fuels which, however, in view of their overexploitation in many locations, are ceasing to be effective, in turn causing the need to search for their replacement, e.g. in the Arctic (Bąk et al. 2021). Moreover, the excessive use of fossil fuels not only leads to the accelerated pace of reducing their reserves, but also has a significant negative impact on the environment, causing increased health risk and a growing threat of a global climate change (global warming) linked to the emission of greenhouse gases (Hall et al. 1991). All this means that the production of energy from non-renewable sources is becoming more expensive and thus less profitable, but—even more importantly—it brings negative effects not just for the environment, as well as affects the entire economy and social sphere. Therefore, developing renewable sources of energy is being treated as the antidote to the problems of the degradation of the natural environment. Many countries, including those in the EU, pledge to limit the mining and exploitation of emission deposits, replacing them with renewable sources of energy (RSE). These are considered as clean sources of energy, and their optimal exploitation minimises impact on the environment, produces minimum waste and is sustainable on the basis of both present and future socio-economic needs (Temiz and Gokmen 2010; Panwar et al. 2011; Ackermann et al. 2017). Additionally, renewable energy technologies provide an ideal opportunity for reducing emissions of greenhouse gases and limiting global warming by replacing the conventional sources of energy. This issue has been noted and widely discussed internationally. According to Ahlstrom et al. (2015), renewable energy is quickly becoming a driving force of new markets and changes in the market structure. The prevention of an energy crisis constitutes one of the most important issues of the twenty-first century. Universal access to affordable and clean energy will be of crucial importance for future achievements of humankind as well as raising global living standards (Veers et al. 2019). The International Energy Agency (IEA 2022) declares that renewable sources of energy were the only ones which in 2020 showed an increase in demand despite the pandemic, whereas the use of all other fuels declined. It is expected in coming years that there will be new global growth in energy generated from renewable resources. The law is the fundamental instrument in providing energy security. In the European Union, the issues of environmental protection are increasingly gaining priority, and laws concerning the protection of the environment have become one of the fastest developing sectors of European law.

2 The Green Economy and Renewable Sources of Energy

The concept of a green economy is focused on perceiving threats resulting from the expansive socio-economic activity of humankind, irreversibly destroying the natural environment and its limited resources. The term 'green economy' was first used in the pioneering report prepared for the British government in 1989, *Blueprint for a Green Economy*, by a group of leading environmentalist economists (Pearce et al. 1989). In 2008, the term was used again in the context of the discussion on the financial crisis and concerns about global recession. The UNEP—the UN agenda on the protection of the environment—supported the idea of 'green stimulation packages' and identified specific areas in which large-scale public investment could stimulate 'green economy' (Atkisson 2012). In June 2009, as part of the preparations for the UN conference on climate change in Copenhagen, the UN published a declaration in support of a green economy as a transformation aimed at resolving several crises. The declaration expressed a conviction that a green economy may contribute to an economic revival as well as decrease the dangers linked to crises regarding food, water, energy, ecosystems and climate. In February 2010, ministers and diplomats heading delegations to the Global Ministerial Environment Forum of UNEP in Nusa Dua, confirmed in their joint declaration that the concept of a green economy "*can to a large degree face the present challenges and ensure possibilities of economic growth, as well as many advantages for all nations*". In March 2010, the General Assembly agreed that a green economy in the context of the sustainable development and elimination of poverty would be one of the two leading subjects in Rio + 20 (resolution 64/236). This drew international attention to the green economy and the related concepts, and also to the appearance of numerous publications and reports connected with them. One of the key reports was the *Green Economy Report* published by UNEP in November 2011. It contains, among others, a working definition of a green economy, quoted in many publications. In recent years, numerous non-governmental organisations and partnerships have also emerged, focused on promoting the concept of a green economy, and carrying out research, analyses and informational activities. At present, there is no internationally approved definition of a green economy, and the latest publications have proposed at least eight separate variants. For instance, UNEP defined a green economy as "*an economy which leads to the improvement in people's well-being and social equality, and at the same time significantly reduces environmental threats and ecological shortcomings. It produces low-emission, saves resources and favours social integration*" (UNEP 2011a). The document entitled *Towards a Green Economy: Pathways to Sustainable Development and Poverty Eradication*" (UNEP 2011b) described in detail stages of implementing a green economy through defining green investment which targets mainly the supply and sustainable use of natural capital and energy. In Europe, the role of the main initiator of actions aimed at transforming a traditional economy into an economy socially responsible and integrated with the natural environment, has been played by the European Commission. The key documents in propagating the idea of a green economy are *Lisbon Strategy*, *Green Paper* and *White Paper*.

Currently, the leading document from the viewpoint of a green economy is the strategy *Europa 2020*, whose core is the support for sustainable development in all areas of the economy, the environment and societal life. One of its leading projects is the document entitled *Roadmap to a Resource-Efficient Europe*, which defines goals and principles of the transformation into a resource-efficient and low-emissions economy, followed by the more specific, detailing tasks and deadlines, *Roadmap to the transformation into a competitive low-emissions economy by 2050*".

To monitor the status of a green economy, indicators from four main subject areas are typically used, such as (Daniek 2020):

1. the natural capital—comprising indicators describing the condition of natural environment,
2. environmental effectiveness of production—this group comprises indicators reflecting links between the natural environment and the economy,
3. environmental quality of life of the population—presented by indicators serving to monitor links between the natural environment and society,
4. economic policies and their consequences—comprising indicators characterising the instruments which impact on the economy and society, creating the desired directions of development aimed at 'greening' the economy.

From the viewpoint of the considerations contained in this chapter, it is necessary to look closer at the second group of indicators referring to the implementation of natural resources, the labour and capital in the production processes, in order to make products and services. The indicators from that group usually include measures describing the use of energy, renewable and non-renewable resources and issues connected with the emission of greenhouse gases as negative effects of human activity. Energy is utilised both in production processes and in households. Its effective use in the economy constitutes an important factor affecting the level of production costs and the competitiveness of products on the international market, while the non-rational use leads to problems connected with environmental pollution (through emissions of greenhouse gases) and to the reduction of the available resources of energy. Demand for energy is constantly on the rise, hence among the main priorities of a green economy one should mention, among others, an improvement in energy efficiency and the rational use of the existing energy resources. An alternative for the traditional primary non-renewable sources of energy (fossil fuels) can be provided by renewable sources of energy (RSE).

Growing interest in RSE was first observed at the end of the 1990s, caused by the rise in crude oil prices, which—besides the serious financial consequences—also exposed the dependence of world economies on fossil fuels (Abbasi et al. 2011). It was a little later that the new terminology emerged, expressions such as 'green energy', 'clean energy', 'energy from ecological sources', and 'renewable energy'. The word 'green' brings to mind a world without pollution and environment-friendly. This is why 'green energy' reflects the idea of generating energy from natural resources, such as sunlight, wind, rain, tides, plants, algae, geothermal energy, etc., with no or slight impact on the environment. These energy resources are renewable, which means that they are replenished in a natural way (Kalyani et al. 2015).

In the subject literature, the term green energy most frequently refers to electric power generated from renewable sources of energy (Marks-Bielska et al. 2020; Kumar and Majid 2020). Many countries do not have legally approved definitions of this term nor guarantee its protection. In order to remedy the lack of clarity in its definitions, countries have introduced certificates, guarantees of origin, labels or quality signs which identify ecological electric power. Every European country has its own register of Guaranteed Origin, which owing to the international coordination, can be the subject of transborder trading. The certificates used to prove the origins of energy in Europe are those from the European System of Energy Certificates (EECS-GoO), which in 2016 replaced the earlier RECS (Renewable Energy Certificates System).

According to Jianzhong et al. (2018), renewable sources of energy are responsible for much lower emission of greenhouse gases, in particular CO_2 and other pollutants, and actually contribute to a significant reduction in their emission. Moreover, the production of renewable energy has minimal impact on the natural environment (Jacobson 2009), as well as lower infrastructural requirements compared to coal power plants. The decentralised production of renewable energy allows to satisfy the demand for energy in rural areas, remote and sparsely populated territories, including deserts and mountain zones, nature reserves and special-protection areas. Additionally, the realisation of projects of generating renewable energy affects social and territorial development, in particular in rural areas where it can create jobs and bring other economic benefits, while entities producing renewable energy generate lower costs of maintenance compared with traditional power plants.

Increasing the share of energy from renewable sources in the total gross use of energy constitutes an element of EU energy policies. The EU legislature regarding the promotion of renewable sources of energy has significantly evolved in recent years. In 2009, EU leaders decided that by 2020 20% of energy used in the EU has to come from renewable sources. In 2018, the target was set for 2030: by then 32% of energy used in the EU has to be generated from renewable sources. In July 2021, owing to a new agenda of the EU regarding climate, it was proposed to co-legislators to amend the 2030 target to 40%. Talks are also underway in relation to future policy framework in the period after 2030.

3 The Essence and Kinds of Renewable Sources of Energy

Renewable Sources of Energy (RSE) are natural, recurrent processes which allow to generate energy as an alternative to fossil fuels. The most important of those are (Panwar et al. 2011):

- solar energy
- wind energy
- water energy
- geothermal energy
- biomass energy.

Solar energy is a very rich resource which will never be exhausted and can be stored in many different ways world-wide. It can supply over 1000 times more than the global demand, yet currently a mere 0.02% of its full potential is being used (Devanbhaktuni et al. 2013). It is beneficial, especially for developing countries (Chua and Oh 2012), due to, among others, the fact that the majority of them are located in regions with optimum access to sunlight, and solar power systems are relatively inexpensive and universally applicable (Devanbhaktuni et al. 2013). Solar power can be transformed into electricity via application of diverse technologies such as photovoltaic panels (PV), concentration of solar power (CSP) and concentration of photovoltaics (CVT).

Wind energy has been developing at a very fast pace (Chen 2021), in some countries becoming the main source of power, e.g. in Denmark around 48% of the electricity consumed in 2020 was generated by wind power (Statista 2022). Wind technology transforms the power created by wind into electric or mechanical energy using wind turbines (Balat 2005). For developing countries such as India, wind turbines provide an attractive source of energy production (Singh et al. 2004), which can be installed and transmitted very quickly, even in remote, and not easily accessible mountainous areas. Electric power generated from wind is unlimited, and can mean saving a few billion barrels of crude oil, as well as avoiding the production of many million tonnes of carbon dioxide (Jianzhong et al. 2018). According to the International Renewable Energy Agency (IRENA 2021), the average cost of wind power is decreasing, becoming comparable with the technology of producing energy based on fossil fuels.

Water energy. Hydroelectric power plants have been used for decades to produce relatively inexpensive renewable energy but these systems are subject to the limitations of natural rainfall and geographical topology. The majority of locations worldwide suitable for creating large hydroelectric resources have already been put into use (Kroposki et al. 2017). For example, Iceland satisfies 100% of its demand for electricity with the use of geothermal and water energy (Kroposki et al. 2017). Other countries whose electricity supplies rely on a large share of renewable energy based on water power include Norway (97%), Costa Rica (93%), Brazil (76%) and Canada (62%). Another source of renewable energy with enormous potential is provided by seas and oceans. Sea/ocean power technologies employ energy produced by tides, the constant movement of waves as well as the natural differences in temperature and levels of salt in seawater. Most importantly, their use does not require supply of fuel, and does not generate pollution. The application of larger amounts of ocean power additionally stimulates other no-emissions renewable sources of energy such as wind and solar power. According to Mattiazzo (2019), technology based on the power of sea waves is still being developed due to the relatively small number of the sold equipment, most of which at prototype level.

Geothermal energy originates from the interior of the Earth (Barbier 2002). It has already been used for decades to produce electricity, both for heating interiors and for industrial processes. It is considered to be a RSE because of the unlimited heat emanating from inside of our planet. The energy is extracted mainly from the heat of water, steam and hot dry rocks. The most productive deposits are located deep under the Earth's crust, which makes geothermal power the most difficult RSE to exploit. There are certain regions which are particularly attractive for the production of geothermal power, usually the thermically active areas in the Earth's crust, close to the edges of tectonic plates (Salazar et al. 2017). The global potential of producing geothermal power amounted to 15,608 MW at the end of 2020. The largest producers of this type of energy are: the USA, Indonesia, the Philippines, Turkey, New Zealand, Mexico, Italy, Kenya, Iceland and Japan (ThinkGeoEnergy 2021).

Biomass energy. Biomass is the vegetable and animal matter undergoing the process of biodegradation. The abundance of biomass and its huge potential as a renewable source of energy makes it a suitable alternative to use in the production, conversion and storage of energy (Kumar et al. 2020). It is used to produce biofuels—gas, liquid and solid (pellets and briquettes). The most effective way of obtaining energy from biomass is through burning it, and the heat created in this way is used to produce RSE power from a renewable source, is self-generated and independent of weather conditions. Biomass in solid form is generally used to heat homes, using special boilers in which the prepared material is burned. The most suitable resources for producing biofuels include oil crops and plants with high sugar content.

Biogas production is a versatile source of renewable energy, since methane can be used as a substitute for fossil fuels both to generate heat and power, as well as vehicle fuel (Weiland 2010). This makes it far more attractive than other forms of producing bioenergy, and was judged to be one of the most energy-efficient and environment-friendly production technologies (Hall et al. 1991). Biogas can be obtained from different materials, applying various methods of fermentation (Borjesson and Berglund 2006).

RSE, due to the fact that they comprise five different kinds of energy, constitute a very wide and multi-dimensional research trend. Pan and Wang (2021) showed that every kind of renewable energy has its own specificity and related threats, which should be closely examined by all decision-makers prior to deciding about investment in any specific kind of RSE.

A search in the Web of Science identified over 122 thousand publications concerning solar energy, over 37 thousand regarding wind energy, around a thousand connected with hydro energy, circa 7 thousand on geothermal energy, and over 7 thousand on biomass energy. Table 1 presents a review of the current research directions regarding individual kinds of renewable energy, most of the publications date from this year (2022).

4 Main Directions of Green Energy Development

The increasing demand for energy, in line with the growing global population and climate change, require urgent investment in sustainable energy (Vargasa et al. 2022). The modernisation of global production of energy aimed at gradual abandoning of hydrocarbons in favour of environment-friendly renewable sources of energy, as a guarantee of safety for humankind, has been present in scientific discourse for many

Table 1 Research directions in the field of types of renewable energy

Directions of research	Authors
Solar energy	
Designs, storage and cost-effectiveness of photovoltaic systems Types of photovoltaic systems Storing solar power Efficiency of solar power systems	Formolli et al. (2022), Legner nad Femenias (2022), Cubukcu and Tari (2022) Pandey et al. (2022), Houchati et al. (2022) Hillers-Bendtsen et al. (2022), Wattana and Aungyut (2022) Jiang et al. (2022), El Alani et al. (2022), Walshe et al. (2022), Preethi (2022), Chen et al. (2022), Gamal et al. (2022)
Wind energy	
Cost of installing and cost-effectiveness Difficulties in installing and functioning of wind power plant Impact on the environment	Millstein et al. (2022), Hemmati et al. (2022), Al-Mhairat and Al-Quraan (2022), Abdelsattar et al. (2022) Liu et al. (2022), Safari et al. (2022), Haydon et al. (2022) Horswill et al. (2022) Zalhaf et al. (2022), Muoneke et al. (2020), Azzam et al. (2022) Radtke et al. (2022)
Hydro energy	
Cost of installing and cost-efficiency Difficulties in installing and functioning of hydro power plant Impact on the environment	Kim et al. (2022), Pathak and Khatod (2022), Li et al. (2022), Haas et al. (2022), Morabito et al. (2022) Emmanouil et al. (2022) Isiksal (2022), Saraswat and Digalwar (2022), Mohapatra (2022)
Geothermal energy	
Cost of installing and cost-effectiveness Barriers and incentives for further improvements Impact on the environment and sustainable development	Birdsell et al. (2021), Alsagri et al. (2021), Kurnia et al. (2021) Zhai et al. (2021), Jia et al. (2021) Soltani et al. (2021), Shahzad et al. (2021), Canbaz et al. (2022)
Biomass energy	
Cost of installing and cost-effectiveness Impact on the environment and sustainable development	Liu et al. (2021);,Wu et al. (2021), Zhang et al. (2021) Jiang et al. (2021), Feng et al. (2021), Long et al. (2021)

years. Radchenko et al. (2021), when considering, among others, possibilities of producing energy, indicated four types of energy security:

1. Model of guaranteed energy security (typical of countries that are not only able to fully meet their own energy demand, but also have significant reserves for sale: the UAE, Venezuela, Russia, other OPEC countries);
2. Model of self-sufficient energy security (typical of countries whose own energy resources are sufficient for domestic needs, but not enough for exports, e.g. India);

3. Model of insufficient energy security (typical of countries whose energy needs require importing energy and measures aimed at diversifying energy sources, stimulating the introduction of alternative energy technologies. Such countries include the EU, Ukraine, etc.);
4. Model of crisis energy security (when a country is forced to import more than 50% of the required amount of energy, e.g. Japan).

Hence it is not surprising that the level of investment in renewable energy production in the EU-28 is constantly growing. Governments and businesses around the world have pledged to add approximately 826 GW of non-hydro renewable power in the course of this decade, up to 2030 (FS-UNEP 2020). In 2019, the amount of new renewable power capacity added (excluding large hydro power systems) was the highest ever, at 184 GW, 20 GW more than in 2018. This included 118 GW of new solar systems, and 61 GW of wind turbines.

In 2022, the annual global increase of renewable energy went up by 35%, up to approximately 280 GW (see Fig. 1), with the greatest increase year on year since 1999 noted in 2020 (by 45%). The growing tendency, although not so significant, has been also observed in European countries (see Fig. 2).

Renewable energy is able to satisfy two-thirds of the total global energy demand and contribute to the reduction in emission of greenhouse gases which is needed from now up to 2050 in order to limit the growth of the average global temperature of the Earth surface below 2 °C (Gielen et al. 2019). Ensuring the final elimination of carbon dioxide emissions will require new technologies and innovation, especially in the transport and production sectors.

It is expected (IEA 2022), that the greatest increase in RSE production capacity will take place in Germany, followed by France, the Netherlands, Spain, Great Britain and Turkey. This strong growth results from multiple countries extending their policies to meet the EU 2030 climate target, and from the corporate power purchase agreement (PPA) markets booming in several countries:

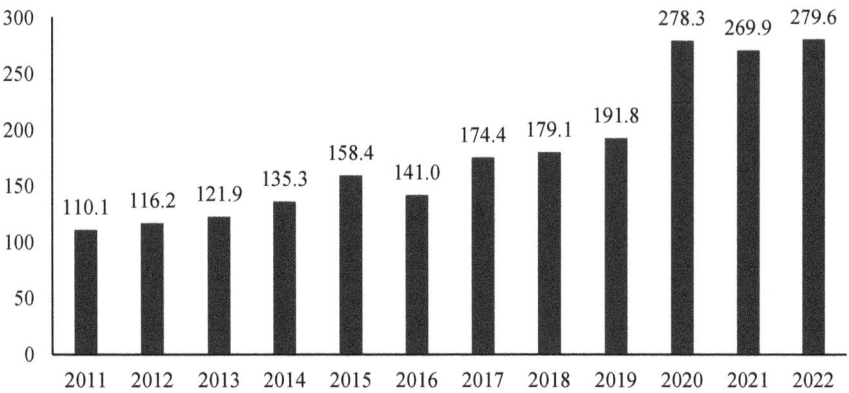

Fig. 1 Increase in renewable energy globally in 2011–2022 (in GW). *Source* IEA (2022)

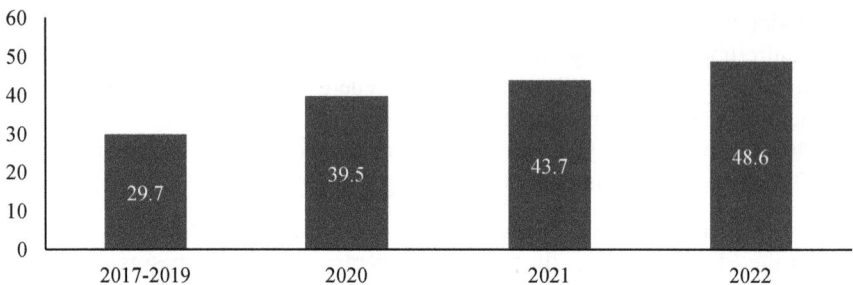

Fig. 2 Annual increases of renewable energy in Europe in 2017–2022 (in GW). *Source* IEA (2022)

- Germany—support for solar PV, wind and bioenergy with higher auction volumes through Germany's Renewable Energy Act 2021 (EEG);
- the Netherlands—allocation of the new SDE++ scheme in December 2020;
- Turkey—extension of feed-in tariff (FIT) scheme for all renewables;
- Poland—new auction awarded almost 1 GW of PV in December 2020;
- Spain—record corporate PPA agreements signed in 2020;
- Sweden—low wind generation costs stimulate a boom in the corporate PPA market; continuation of the PV rebate programme;
- the United Kingdom—proposal to re-include onshore wind and solar PV in the 2021 contracts for difference (CfD) auction.

The study by Bąk et al. (2021) identified five main factors influencing the development of RSE in EU countries, namely:

1. political factor—allows to check whether there have been changes in the policies that support the development of RSE use, including an assessment of their effectiveness;
2. legal factor—allows to introduce changes aimed at increasing the effectiveness of activities, including policies;
3. geographical (natural) environment factor—allows to monitor efficiency in the indicated groups, assess the jump of changes and the direction of change of groups into which the countries were classified;
4. financial factor—allows to assess whether the level of financing policies was adequate;
5. information and education factor—reflects the need to finance the information policy, promotional activities, and sharing good practices, including the effectiveness of reporting on ESG factors related to RSE.

Renewable energy is of key importance for the success of the Agenda 2030 approved by the UN General Assembly in 2015. The global target in respect of energy—SDG 7—comprises three main goals: ensuring inexpensive, reliable and universal access to modern power services; significant increase in the share of renewable energy in the global energy supply basket; doubling the global pace of improving energy efficiency. In order to reach the aims of sustainable growth, it is necessary to achieve a global energy transition, which is possible owing to technological innovations in the field of renewable energy.

The latest data published in April 2022 by the International Renewable Energy Agency (IRENA 2022) show that the production of renewable energy is continuously growing and gaining pace. At the end of 2021, global renewable generation capacity amounted to 3064 Gigawatt (GW), increasing the stock of renewable power by 9.1%. Although hydropower accounted for the largest share of the global total renewable generation capacity with 1230 GW, IRENA's *Renewable Capacity Statistics 2022* shows that solar and wind continued to dominate new generating capacity. Both technologies together contributed 88% to the share of all the new renewable capacity in 2021. Solar capacity led with 19% increase, followed by wind energy, which increased its generating capacity by 13%.

5 Summary

In line with the global tendency, Europe is also moving towards renewable sources of energy, whilst the solutions applied in this respect depend on local conditions, such as large amounts of sunlight (photovoltaic installations) and wind power (wind turbines). Independently of the selected type of RSE, individual countries around the world, including the EU, are increasing the share of renewable energy sources in the general balance of their energy production. The abandonment of fossil fuels has already become a global trend. The growing share of RSE contributes to saving natural resources, and to improving the condition of the natural environment through reducing the emission of pollutants into the environment and cutting down the amounts of the produced waste. Moreover, it contributes to increasing the energy security of countries, and also to improving electricity supplies to less urbanised areas. According to EU assumptions, alternative sources of energy will constitute in future a significant part of the energy balance in Europe.

References

Abbasi T, Premalatha M, Abbasi S (2011) The return to renewables: will it help in global warming control? Renew Sustain Energy Rev 15(1):891–894. https://doi.org/10.1016/j.rser.2010.09.048

Abdelsattar M, Arafa HW, Elbaset A, Kamel S, Kasem HAA, Khan B, Zaki DA (2022) Voltage stability improvement of an Egyptian power grid-based wind energy system using STATCOM. Wind Energy 25(6):1077–1120. https://doi.org/10.1002/we.2716

Ackermann T, Prevost T, Vittal V, Roscoe AJ, Matevosyan J et al (2017) Paving the way: a future without inertia is closer than you think. IEEE Power Energy Mag 15(6):61–69. https://doi.org/10.1109/MPE.2017.2729138

Ahlstrom M, Ela E, Riesz J, O'Sullivan J, Hobbs BF et al (2015) The evolution of the market: designing a market for high levels of variable generation. IEEE Power Energy Mag 13(6):60–66. https://doi.org/10.1109/MPE.2015.2458755

Al-Mhairat B, Al-Quraan A (2022) Assessment of wind energy resources in Jordan using different optimization techniques. Processes 10(1):105. https://doi.org/10.3390/pr10010105

Alsagri AS, Chiasson A, Shahzad MW (2021) Geothermal energy technologies for cooling and refrigeration systems: an overview. Arab J Sci Eng. https://doi.org/10.1007/s13369-021-06296-x

Atkisson A (2012) Life beyond growth. Institute for Studies in Happiness, Economy, and Society. http://sdg.iisd.org/news/report-on-life-beyond-growth-proposes-combining-green-economy-and-national-happiness-concepts/. Accessed 09 March 2022

Azzam SM, Sleem MM, Sallam KM, Munasinghe K, Abohany AA (2022) A framework for evaluating sustainable renewable energy sources under uncertain conditions: a case study. Int J Intell Syst 1–34. https://doi.org/10.1002/int.22858

Bąk I, Spoz A, Zioło M, Dylewski M (2021) Dynamic analysis of the similarity of objects in research on the use of renewable energy resources in European Union countries. Energies 14:3952. https://doi.org/10.3390/en14133952

Balat M (2005) Usage of energy sources and environmental problems. Energy Explor Exploit 23(2):141–168

Barbier E (2002) Geothermal energy technology and current status: an overview. Renew Sustain Energy Rev 6(1–2):3–65. https://doi.org/10.1016/S1364-0321(02)00002-3

Birdsell DT, Adams BM, Saar MO (2021) Minimum transmissivity and optimal well spacing and flow rate for high-temperature aquifer thermal energy storage. Appl Energy 289:116658. https://doi.org/10.1016/j.apenergy.2021.116658

Borjesson P, Berglund M (2006) Environmental systems analysis of biogas systems. Part I. Fuel-cycle emissions. Biomass Bioenergy 30(5):469–485. https://doi.org/10.1016/j.biombioe.2005.11.014

Canbaz CH, Ekren O, Aksoy N (2022) Review of wellbore flow modelling in CO_2-bearing geothermal reservoirs. Geothermics 98:102284. https://doi.org/10.1016/j.geothermics.2021.102284

Chen Z (2021) Wind power: an important source in energy systems. Wind 1(1):90–91. https://doi.org/10.3390/wind1010006

Chen Y, Xu J, Wang J, Lund P, Wang D (2022) Configuration optimization and selection of a photovoltaic-gas integrated energy system considering renewable energy penetration in power grid. Energy Conv. Manag. 254:115260. https://doi.org/10.1016/j.enconman.2022.115260

Chua SC, Oh TH (2012) Solar energy outlook in Malaysia. Renew Sustain Energy Rev 16(1):564–574. https://doi.org/10.1016/j.rser.2011.08.022

Cubukcu N, Tari I (2022) Buildings sector from a sustainable carbon constrained energy generation perspective. Energy Build. 259:111865. https://doi.org/10.1016/j.enbuild.2022.111865

Daniek K (2020) Green economy indicators as a method of monitoring development in the economic, social and environmental dimensions. Soc Inequalities Econ Growth 62(2):150–173. https://doi.org/10.15584/nsawg.2020.2.10

Devanbhaktuni V, Alam M, Shekara Sreenadh Reddy Depuru S, Green RC, Nims D, Near C (2013) Solar energy: trends and enabling technologies. Renew Sustain Energy Rev 19:555–564. https://doi.org/10.1016/j.rser.2012.11.024

El Alani O, Ghennioui H, Abraim M, Ghennioui A, Blanc P, Saint-Drenan YM, Naimi Z (2022) Solar energy resource assessment using GHI and DNI Satellite Data for Moroccan Climate. In: Saidi R, El Bhiri B, Maleh Y, Mosallam A, Essaaidi M (eds) Advanced technologies for humanity. ICATH 2021. Lecture Notes on Data Engineering and Communications Technologies, vol 110. Springer, Cham. https://doi.org/10.1007/978-3-030-94188-8_26

Emmanouil S, Nikolopoulos EI, Francois B, Brown C, Anagnostou EN (2022) Evaluating existing water supply reservoirs as small-scale pumped hydroelectric storage options—a case study in Connecticut. Energy 226:120354. https://doi.org/10.1016/j.energy.2021.120354

Feng J, Zhu H, Xu Y, Jiang J, Pan H (2021) Preparation and characterization of high-performance activated carbon from papermaking black-liquor at low temperature. J Anal Appl Pyrolysis 159:105292. https://doi.org/10.1016/j.jaap.2021.105292

Formolli M, Croce S, Vettorato D, Paparella R, Scognamiglio A, Mainini AG, Lobaccaro G (2022) Solar energy in urban planning: lesson learned and recommendations from Six Italian case studies. Appl Sci-Basel 12(6):2950. https://doi.org/10.3390/app12062950

FS-UNEP (2020) Global trends in renewable energy investment. https://wedocs.unep.org/bitstream/handle/20.500.11822/32700/GTR20.pdf?sequence=1&isAllowed=y. Accessed 09 March 2022

Gamal A, EL-Ghetany HH, Elsayed AM, Zedan IT (2022) Theoretical analysis of different solar water pumping irrigation systems for seasonal crops in three geographical locations in Egypt. Egypt J Chem 65(3):227–243. https://doi.org/10.21608/ejchem.2021.78302.3834

Gielen D, Boshella F, Saygin D, Bazilian MD, Wagner N, Gorini R (2019) The role of renewable energy in the global energy transformation. Energ Strat Rev 24:38–50. https://doi.org/10.1016/j.esr.2019.01.006

Haas J, Prieto-Miranda L, Ghorbani N, Breyer C (2022) Revisiting the potential of pumped-hydro energy storage: a method to detect economically attractive sites. Renew Energy 181:182–193. https://doi.org/10.1016/j.renene.2021.09.009

Hall DO, Mynick HE, Williams RH (1991) Cooling the greenhouse with bioenergy. Nature 353:11–12

Haydon B, Cole J, Dunn L, Keyantuo P, Chow FK, Moura S, Vermillion Ch (2022) Generalized empirical regret bounds for control of renewable energy systems in spatiotemporally varying environments. J. Dyn. Syst. Meas. Control-Trans. 144(4):44501. https://doi.org/10.1115/1.4052396

Hemmati R, Nosratabadi SM, Mehrjerdi H, Bornapour M (2022) Coordination of thermal/wind energies in power-to-gas process for cost/pollution abatement considering wind energy recovery. Energy Sources Part A-Recovery Util Environ Eff 44(1):632–649. https://doi.org/10.1080/15567036.2022.2049930

Hillers-Bendtsen AE, Kjeldal FO, Hoyer NM, Mikkelsen KV (2022) Optimization of the thermo-chemical properties of the norbornadiene/quadricyclane photochromic couple for solar energy storage using nanoparticles. Phys Chem Chem Phys 24(9):5506–5521. https://doi.org/10.1039/D2CP00226D

Horswill C, Miller JAO, Wood MJ (2022) Impact assessments of wind farms on seabird populations that overlook existing drivers of demographic change should be treated with caution. Conserv. Sci. Pract. 4:e12644. https://doi.org/10.1111/csp2.12644

Houchati M, AbdlMonem BH, Khraisheh M (2022) Predictive modeling for rooftop solar energy throughput: a machine learning-based optimization for building energy demand scheduling. J Energy Resour Technol 144(1):11302. https://doi.org/10.1115/1.4050844

IEA (2022) Renewable energy market update. Outlook for 2021 and 2022. https://iea.blob.core.windows.net/assets/18a6041d-bf13-4667-a4c2-8fc008974008/RenewableEnergyMarketUpdate-Outlookfor2021and2022.pdf. Accessed 09 March 2022

IRENA (2021) Renewable power generation costs in 2020. https://www.irena.org/-/media/Files/IRENA/Agency/Publication/2021/Jun/IRENA_Power_Generation_Costs_2020.pdf. Accessed 09 March 2022

IRENA (2022) Renewables take Lion's share of global power additions in 2021. https://www.irena.org/newsroom/pressreleases/2022/Apr/Renewables-Take-Lions-Share-of-Global-Power-Additions-in-2021. Accessed 22 April 2022

Isiksal AZ (2022) The decline in carbon intensity: the role of financial expansion and hydro-energy. Environ Sci Pollut Res 29(11):16460–16471. https://doi.org/10.1007/s11356-021-16117-1

Jacobson MZ (2009) Review of solutions to global warming, air pollution, and energy security. Energy Environ Sci 2:148–173. https://doi.org/10.1039/B809990C

Jia GS, Ma ZD, Xia ZH, Wang JW, Zhang YP, Jin LW (2021) Investigation of the horizontally-butted borehole heat exchanger based on a semi-analytical method considering groundwater seepage and geothermal gradient. Renew Energy 171:447–461. https://doi.org/10.1016/j.renene.2021.02.129

Jiang L, Liu FY, Yang W, Li C, Zhu B, Zhu X (2021) Production of 1,3-propanediol and lactic acid from crude glycerol by a microbial consortium from intertidal sludge. Biotechnol Lett 43(3):711–717. https://doi.org/10.1007/s10529-020-03063-0

Jiang S, Bai Y, Ma Z, Jin S, Zou Ch, Tan Z (2022) Recent advances of monolithic all-Perovskite Tandem solar cells: from materials to devices. Chin J Chem 40(7):856–871. https://doi.org/10.1007/s10529-020-03063-0

Jianzhong X, Assenova A, Erokhin V (2018) Renewable energy and sustainable development in a resource-abundant country: challenges of wind power generation in Kazakhstan. Sustainability 10(9):3315. https://doi.org/10.3390/su10093315

Kalyani VL, Dudy MK, Pareek S (2015) Green energy: the need of the world. J Manag Eng Inf Technol 2(5):2394–8124

Kim ES, Sun H, Park H, Shin S, Chae EJ, Ouderkirk R, Bernitsas MM (2022) Development of an alternating lift converter utilizing flow-induced oscillations to harness horizontal hydrokinetic energy. Renew Sust Energ Rev 145:111094. https://doi.org/10.1016/j.rser.2021.111094

Kroposki B, Johnsona B, Zhang Y, Geworgian V et al (2017) Achieving a 100% renewable grid: operating electric power systems with extremely high levels of variable renewable energy. IEEE Power Energy Mag. 15(2):61–73. https://doi.org/10.1109/MPE.2016.2637122

Kumar AK, Bhattacharya T, Hasnain SMM, Nayak AK (2020) Applications of biomass-derived materials for energy production, conversion, and storage. Mater Sci Energy Technol 3:905–920. https://doi.org/10.1016/j.mset.2020.10.012

Kumar CHR, Majid MA (2020) Renewable energy for sustainable development in India: current status, future prospects, challenges, employment, and investment opportunities. Energy, Sustain Soc 10(2). https://doi.org/10.1186/s13705-019-0232-1

Kurnia JC, Putra ZA, Muraza O, Ghoreishi-Madiseh SA, Sasmito AP (2021) Numerical evaluation, process design and techno-economic analysis of geothermal energy extraction from abandoned oil wells in Malaysia. Renew Energy 175:868–879. https://doi.org/10.1016/j.renene.2021.05.031

Legner M, Femenias P (2022) The implementation of conservation policy and the application of solar energy technology in small house areas: Stockholm. Sweden. Hist Env-Policy Pract https://doi.org/10.1080/17567505.2022.2048463

Li H, Yao X, Tachega MA, Ahmed D (2022) Path selection for wind power in China: hydrogen production or underground pumped hydro energy storage? J Renew Sustain Energy 13(3):3590. https://doi.org/10.1063/5.0041207

Liu M, Sun Z, Li Q, Wei Z, Liang B (2021) Driving and influencing factors of biomass energy utilization from the perspective of farmers. Int J Heat Technol 39(1):267–274. https://doi.org/10.18280/ijht.390130

Liu Y, Qian Y, Berg LK (2022) Local-thermal-gradient and large-scale-circulation impacts on turbine-height wind speed forecasting over the Columbia River Basin. Wind Energy Sci 7(1):37–51. https://doi.org/10.5194/wes-7-37-2022

Long F, Liu W, Jiang X, Zhai Q, Cao X, Jiang J, Xu J (2021) State-of-the-art technologies for biofuel production from triglycerides: a review. Renew Sust Energ Rev 148:111269. https://doi.org/10.1016/j.rser.2021.111269

Marks-Bielska R, Bielski S, Pik K, Kurowska K (2020) The importance of renewable energy sources in Poland's energy mix. Energies 13:4624. https://doi.org/10.3390/en13184624

Mattiazzo G (2019) State of the art and perspectives of wave energy in the Mediterranean Sea: backstage of ISWEC. Front Energy Res 29. https://doi.org/10.3389/fenrg.2019.00114

Millstein D, Bolinger M, Wiser R (2022) What can surface wind observations tell us about inter-annual variation in wind energy output? Wind Energy 25(6):1142–1150. https://doi.org/10.1002/we.2717

Mohapatra NK (2022) Climate change, energy security and societal vulnerability in Eurasia. J Clim Chang 6(2):1–14. https://doi.org/10.3233/JCC200008

Morabito A, Spriet J, Vagnoni E, Hendrick P (2022) Underground pumped storage hydropower case studies in Belgium: perspectives and challenges. Energies 13(15):4000. https://doi.org/10.3390/en13154000

Muoneke OB, Okere KI, Nwaeze CN (2020) Agriculture, globalization, and ecological footprint: the role of agriculture beyond the tipping point in the Philippines. Environ Sci Pollut Res. https://doi.org/10.1007/s11356-022-19720-y

Pan X, Wang Y (2021) Evaluation of renewable energy sources in China using an interval type-2 fuzzy large-scale group risk evaluation method. Appl Soft Comput 108:107458. https://doi.org/10.1016/j.asoc.2021.107458

Pandey A, Pandey P, Tumuluru JS (2022) Solar energy production in India and commonly used technologies—an overview. Energies 15(2):500. https://doi.org/10.3390/en15020500

Panwar N, Kaushik S, Kothari S (2011) Role of renewable energy sources in environmental protection: a review. Renew Sustain Energy Rev 15(3):1513–1524. https://doi.org/10.1016/j.rser.2010.11.037

Pathak DP, Khatod DK (2022) Development of integrated renewable energy system based on optimal operational strategy and sizing for an un-electrified remote area. IETE J Res. https://doi.org/10.1080/03772063.2021.1939800

Pearce D, Markandya A, Barbier E (1989) Blueprint for a green economy. Routledge, London. https://doi.org/10.4324/9780203097298

Preethi V (2022) Solar hydrogen production in India. Environ Dev Sustain. https://doi.org/10.1007/s10668-022-02157-1

Radchenko O, Radchenko O, Kriukov O, Kovach V, Mykhalchenko O, Abbasov R, Jureniene V (2021) Prospective directions of state regulation of "green" energy development in the context of Ukraine's energy safety. E3S Web Conf 280:09023. https://doi.org/10.1051/e3sconf/202128009023

Radtke J, Yildiz O, Roth L (2022) Does energy community membership change sustainable attitudes and behavioral patterns? Empirical evidence from community wind energy in Germany. Energies 15(3):822. https://doi.org/10.3390/en15030822

Safari MAM, Masseran N, Majid MHA (2022) Wind energy potential assessment using Weibull distribution with various numerical estimation methods: a case study in Mersing and Port Dickson, Malaysia. Theor Appl Climatol 148:1085–1110. https://doi.org/10.1007/s00704-022-03990-0

Salazar SS, Muñoz Y, Ospino A (2017) Analysis of geothermal energy as an alternative source for electricity in Colombia. Geotherm Energy 5(1):27. https://doi.org/10.1186/s40517-017-0084-x

Saraswat S, Digalwar AK (2022) Evaluation of energy sources based on sustainability factors using integrated fuzzy MCDM approach. Int J Energy Sect Manag 15(1):267–290. https://doi.org/10.1108/IJESM-07-2020-0001

Shahzad U, Schneider N, Ben JM (2021) How coal and geothermal energies interact with industrial development and carbon emissions? An autoregressive distributed lags approach to the Philippines. Resour Policy 74:102342. https://doi.org/10.1016/j.resourpol.2021.102342

Singh S, Bhatti TS, Kothari DP (2004) Indian scenario of wind energy: problems and solutions. Energy Sources, Part a: Recovery, Utilization, Environ Effects 26(9):811–819. https://doi.org/10.1080/00908310490465885

Söderholm P (2020) The green economy transition: the challenges of technological change for sustainability. Sustain Earth 3(6). https://doi.org/10.1186/s42055-020-00029-y

Soltani M, Kashkooli FM, Souri M, Rafiei B, Jabarifar M, Gharali K, Nathwani JS (2021) Environmental, economic, and social impacts of geothermal energy systems. Renew Sust Energ Rev 140:110750. https://doi.org/10.1016/j.rser.2021.110750

Statista (2022) Share of wind power coverage in Denmark from 2009 to 2020. https://www.statista.com/statistics/991055/share-of-wind-energy-coverage-in-denmark/. Accessed 7 April 2022

Temiz D, Gokmen A (2010) The importance of renewable energy sources in Turkey. Int J Econ Fin Stud 2(2):23–30

ThinkGeoEnergy (2021) https://www.thinkgeoenergy.com/thinkgeoenergys-top-10-geothermal-countries-2020-installed-power-generation-capacity-mwe/. Accessed 7 April 2022

UNEP (2011a) Annual Report 2011a. https://www.unep.org/resources/annual-report/unep-2011a-annual-report. Accessed 7 April 2022

UNEP (2011b) Towards a green economy: pathways to sustainable development and poverty eradication. https://sustainabledevelopment.un.org/index.php?page=view&type=400&nr=126&menu=35. Accessed 7 April 2022

Van Bavel J (2013) The world population explosion: causes, backgrounds and—projections for the future. Facts, views & vision in ObGyn 5(4):281–291. https://www.ncbi.nlm.nih.gov/pmc/articles/PMC3987379/. Accessed 09 March 2022

Vargasa CA, Caracciolo L, Ball PJ (2022) Geothermal energy as a means to decarbonize the energy mix of megacities. Commun Earth Environ 3(66). https://doi.org/10.1038/s43247-022-00386-w

Veers P, Dykes K, Lantz E, Barth S, Bottasso C et al (2019) Grand challenges in the science of wind energy. Science 366(6464). https://doi.org/10.1126/science.aau2027

Walshe J, Doran J, Amarandei G (2022) Evaluation of the potential of nanofluids containing different Ag nanoparticle size distributions for enhanced solar energy conversion in hybrid photovoltaic-thermal (PVT) applications. Nano Express 3(1):15001. https://doi.org/10.1088/2632-959X/ac49f2

Wattana, B; Aungyut, P. (2022). Impacts of solar electricity generation on the Thai Electricity Industry. Int J Renew Energy Dev-IJRED 11(1):157–163. https://doi.org/10.14710/ijred.2022.41059

Weiland P (2010) Biogas production: current state and perspectives. Appl Microbiol Biotechnol 85(4):849–860. https://doi.org/10.1007/s00253-009-2246-7

Wu D, Chen C, Guo Q, Wang A, Sun K, Jiang J (2021) Facile and green one-step synthesis of Ni3S2@CN carbon nanosheets from sodium lignosulfonate for a supercapacitor electrode. Sustain Energ Fuels 5(19):4895–4903. https://doi.org/10.1039/d1se00946j

Zalhaf AS, Elboshy B, Kotb M, Han Y, Almaliki AH, Aly RMH, Elkadeem MR (2022) A high-resolution wind farms suitability mapping using GIS and Fuzzy AHP approach: a national-level case study in Sudan. Sustainability 14(1):358. https://doi.org/10.3390/su14010358

Zhai Y, Zhang T, Tan X, Wang G, Duan L, Shi Q, Ji C, Bai Y, Shen X, Meng J, Hong J (2021) Environmental impact assessment of ground source heat pump system for heating and cooling: a case study in China. Int J Life Cycle Assess 27:395–408. https://doi.org/10.1007/s11367-022-02034-z

Zhang C, Yang L, Huo S, Su Y, Zhang Y (2021) Optimization of the cell immobilization-based chain-elongation process for efficient n-Caproate production. ACS Sustain Chem Eng 9(11):4014–4023. https://doi.org/10.1021/acssuschemeng.0c07281

Green Energy in the Political Debate

Agnieszka Malkowska and Arkadiusz Malkowski

Abstract In the twenty-first century, energy resources and production are the basis for efficient operation of the world's economies. This is why development of the energy sector has become such an important political, economic and social aspect of every country. It guarantees economic development of the country, as well as maintaining or even enhancing the population's standard of living. This process is accompanied by growing awareness of limited resource availability and the environmental impact of predatory resource management. The use of green energy in the economy is currently one of the major topics of political and social debate, both at the global level and in individual integration formations or countries. The outcome of this debate seems to be an inevitable drift towards climate neutrality. The chapter outlines evolution of policies on the use of energy resources and the major factors influencing the development of green transition policies.

1 Introduction

The last 250 years have been a period of increasing energy dependence in the society. We produce more and consume more. We need more cheap energy from reliable sources. It is the problem of cheap alternatives to traditional energy sources and ensuring energy security both at present and in the future that has dominated the political and social debate throughout the world for many years now.

Energy transition, meaning gradual withdrawal from the use of fossil fuels until total elimination thereof, is presented in the political debate as both necessary and impossible. It is necessary because we, as humanity, should strive to avoid a climatic

---

A. Malkowska
Faculty of Economics, Finance and Management, University of Szczecin, Szczecin, Poland
e-mail: agnieszka.malkowska@usz.edu.pl

A. Malkowski (✉)
Faculty of Economics, West Pomeranian University of Technology in Szczecin, Szczecin, Poland
e-mail: amalkowski@zut.edu.pl

catastrophe. It seems impossible because the current socio-economic system upon which we are fully dependent is based on cheap and accessible energy from fossil fuels.

Presented here, the inconsistency in the approach towards energy transition is the consequence of a number of conflicting interests of individual social groups, countries or integration formations. It is driven by different levels of socio-economic development, different social awareness, different political goals and a number of other considerations that may or may not be expressed in the ongoing political debate on fundamental issues related to the use of green energy. This includes the unavoidable impact of energy transition on the economics and quality of life of the population. The relationship between economics and politics is particularly strong in this case, as Buchanan and Tullock point out by proving that individualistic market logic is inexorably applied to political decision making (Buchanan et al. 1965).

Despite a multi-faceted and lively discussion over the years, we still do not know the answers to the following key questions:

1. What should the energy transition path look like?
2. How to convince and build support for shifting economies and societies to using green energy?
3. How to convince all countries of the world to adopt an economic model based on the use of renewable energy, and can one even talk about a single model in this case?
4. What should energy solidarity of developed countries towards developing ones look like?

Considering the achievements of the political debate on energy transition to date, it seems impossible to convince citizens to turn to renewable energy simply by resorting to slogans on the need to protect the climate or to comply with imposed standards and restrictions. In particular, this concerns meeting the technological requirements imposed by highly developed countries onto those countries that are developing their economies on the basis of conventional energy sources. To achieve this, various activities must be undertaken on many levels. One example of this type of activity in the international, national, regional and local space is conducting an extremely difficult political debate on both the challenges crucial to energy transition and solving the problems antagonising local communities, which significantly influence the shaping of public awareness.

2 Energy Transition in the Public Debate

The first industrial revolution marked a shift away from renewable energy sources (RES). The human race abandoned the wind driving windmills and the energy of falling water, used in many ways, in favour of coal and, in subsequent years, oil. This was mainly due to economic reasons. Replacing wood with coal made it possible to make more profit. Numerous studies indicate that the introduction of steam engines allowed independence from natural forces and revolutionised the entire economy. This enabled the development of mass production and improved the quality of life on a global scale (Pasten and Santamarina 2012; Koc and Teker 2019).

Increasing dependence of developed societies on non-renewable energy resources, including oil, led to a serious global economic crisis in the 1970s. As it turned out, the strategy of unlimited development had one key limitation, namely the access to cheap energy, which was pointed out in their works by, among others, Nordhaus, Coyle and others (Nordhaus 1980; Coyle and Simmons 2014). It was the spectre of economic crash, not growing environmental awareness, that sparked an international political debate on the need for energy transition.

One consequence of the 1970s oil shock was technological progress in the search for new energy sources. A certain symbol of this was installation of 32 solar collectors for heating water on the roof of the White House in 1979, during the presidency of Jimmy Carter. However, this does not negate the fact that the early economic boom is predominantly based on energy-intensive technologies, unfortunately, much of the energy needed by the world economy still comes from non-renewable sources. This is especially true of coal and gas—see Fig. 1.

Despite many declarations, coal and gas are still the dominant sources of energy generation worldwide.

In the face of growing pressure from the international community to take decisive steps to combat climate change and reduce environmental anthropopressure, a political debate has been ongoing for years. It is as complex and turbulent as the problem it addresses. Its outcomes are difficult to predict, as there is still no complete consensus as to whether civilisational development is the cause of the changes taking place or whether they are a natural phenomenon that has been known on Earth for

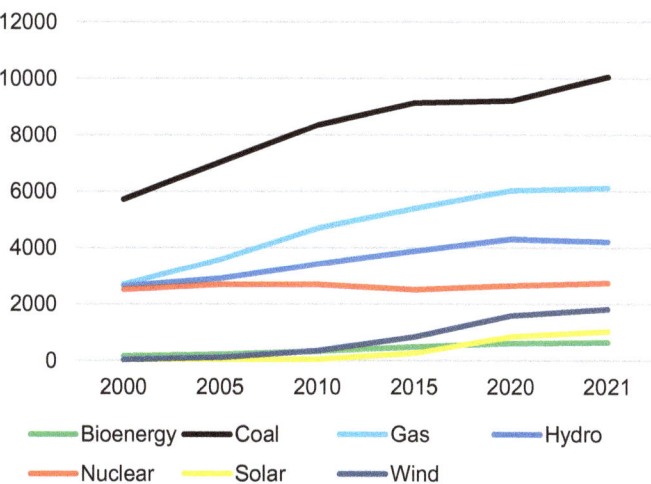

Fig. 1 World electricity generation by source [in Terawatt hours]. *Source* Own compilation based on data from Ember https://ember-climate.org

thousands of years, as pointed out by the Nongovernmental International Panel on Climate Change (NIPCC) in its papers (Singer and Idso 2009). According to Singer, it is the nature, not man, that has decisive influence on climate change (Singer 2008).

Climate processes arouse a lot of emotion, particularly evident in the political, economic and social discourse conducted with vigour since the 1970s. This has resulted in the term "global warming" being one of the key issues of the international debate these days (Kreibich 2007; Gomółka 2010), the object of interest not only of mainstream media, but also of science and politics.

In the face of rapid population growth and shrinking natural resources, the number of international environmental protection regimes is clearly on the rise, which is increasingly shaping the policy on the local, regional and international scale. As a result, the debate around green transition is gaining momentum.

The authors' aim is not to provide a chronological account of how the political discourse on energy transition and widely understood approach to the use of resources has developed. Therefore, the focus is on selected events that, in the authors' opinion, had greatest influence on the direction of social and economic changes promoted nowadays. Attention is paid to those that were the effect of activities of international political organisations (the UN and its agencies), organisations of a political and economic nature (e.g. the European Union), or other entities (e.g. Club of Rome), which significantly influenced activities in the field of economic and energy transition.

In many aspects, the end of the Second World War marked the end of the modern era of international relations and created a new world, whose foundations were to be laid in cooperation within the framework of the United Nations and specialised bodies of international law, such as the International Trade Organisation, the International Monetary Fund and the World Bank. The new political and economic order, which aimed to maintain peace by, among other things, intensifying socio-economic development, was outlined as part of the Bretton Woods Conference of July 1944 (Dooley et al. 2004; Nurzyńska 2017). The political-economic determinants of the Bretton Woods system stemmed primarily from: experience of the Great Depression; radicalisation of societies as a result of economic crash; ineffectiveness of the political system, based on the domination of a few European states; and the desire to create a world economy based on liberal trade. The basis of the consensus developed was a shared belief in capitalism and the need to quickly rebuild the economies destroyed by the war (Hall and Tavlas 2013; Igwe 2018).

The period of two decades after the Second World War was a time that—in international relations—saw not only the establishment of strong organisations in international relations but, above all, the collapse of the previous colonial model, which resulted in the emergence of many new states, particularly in Asia and Africa. Thus, the world of the 1970s was very different from that of the 1940s. There were new challenges and problems to be solved in the new reality. What did not change, however, was that political, economic and military conflicts continued to erupt. The world was clearly divided between the rich highly developed countries, which were in the minority, and the majority of countries which were just building up their economies. It became clear very soon that, despite these divisions, cooperation was necessary. One of the key and first issues that triggered a difficult political debate and, then, cooperation on the global scale were issues connected with the environment and green transition.

3 Role of International Community in the Debate on Green Transition

One of the first international environmental policy initiatives was launched by the United Nations. In July 1968, the Economic and Social Council of the United Nations requested the UN General Assembly in its Resolution 1346 to urgently organise a world conference on the problems of the human environment. In the Resolution (No. 2398) adopted at the same time, the UN General Assembly stressed that scientific and technological development had given rise to unprecedented possibilities of changing and shaping the human environment in accordance with people's needs and aspirations, but that, at the same time, this development posed serious risks if it was not controlled properly. During the aforementioned assembly, UN Secretary General Sithu U Thant delivered his famous speech, calling for seeking alternative ways of human development respecting the natural environment, as well as taking into account the other elements of human reality: social, economic, cultural (Thant 1971; Biermann et al. 2017).

For the time being, the effect of the actions undertaken in the political sphere is the report prepared by UN Secretary General U Thant entitled The problems of human environment (Kozyra et al. 2007). This was the first report in the history of the international community which assumed the need for cooperation in shaping the environmental and climate policy on the global scale (Sohn 1973; Sand 2007; Falkner 2012). Prepared on the basis of opinions and studies, as well as statistical information, the report assessed the condition of the environment and identified the most serious threats occurring locally and nationally, and also regionally and globally. It emphasised that severity of the threats should lead to the identification of a specific hierarchy of their importance and to establishment of a sequence of necessary actions aimed at preventing them and mitigating their negative consequences. The major threats included: reduction in the area of arable land and land for economic use, lack of adequate plans for urban development, disappearance of certain plant and animal species and environmental pollution, especially of soil, water and air.

The report by U Thant was important for development of the international political debate, not only in the field of environmental protection, but also in terms of a broad discussion on the need to revise the existing economic model. It initiated a new process in the international space, involving inclusion of such notions as: environment, sustainable development, ecological security, energy transition, climate responsibility, etc. into international debates.

The venue for a lively debate on the theses of the U Thant Report was the UN Stockholm Conference held on 5-16.06.1972. It was attended by representatives of

113 countries, delegations of governmental organisations and numerous observers from non-governmental organisations. The conference was held under the motto "We only have one Earth". During the conference, environmental protection was elevated to the status of a fundamental state function (Sohn 1973; Handl 2012). Also, a new term emerged in political discourse: "environmental protection policy", which indicated the need to make environmental protection activities a part of state policy (Hultman and Pulé 2018; Katz-Rosene and Paterson 2018). The conference highlighted the need to set up a specialised agency within the UN to deal with environmental protection issues. This led to establishment of the United Nations Environment Programme (UNEP). The purpose of this specialised UN agency was to carry out environmental protection activities and to continuously monitor the condition of the environment worldwide (Conca 1995; Young et al. 2008).

The final declaration of the conference stressed the need to act wisely when taking any action so as not to cause serious and irreversible damage to the environment. Protecting the environment for present and future generations is a goal for all mankind, to be pursued by individuals, communities, businesses and institutions. Nations and international organisations should be involved in environmental protection. It was stressed that environmental protection was vital for economic development of the world and should be treated as a duty of all governments. The need to adopt a common concept and principles that would inspire and guide actions aimed at preserving and improving the human environment was advocated. The conference called for the development of cooperation to preserve and improve the environment. This should be considered, as the foundation for further policy debate on sustainable development and energy transition (Kasperson and Dow 1991; Conca 1995).

The Report of the UN World Commission on Environment and Development, chaired by Gro Harlem Brundtland (Sneddon et al. 2006), proved to be important for developing the policy debate on the need for rational use of the environment. Published in 1987, the Report was entitled "Our Common Future". Its authors pointed to the need to create a fully sustainable model of life, understood as improving the quality of life of people around the world, without predatory exploitation of the earth's natural resources, which met with broad interest from politicians and scholars (see: Pearce and Atkinson 1998; Barkemeyer et al. 2014). At the same time, it was pointed out that this required differentiated actions in individual regions of the world. Above all, integration of actions in three key areas is essential (Brundtland 1989).

Economic growth and equitable distribution of benefits. This is intended to deliver responsible, long-term growth for all nations and communities.

Conservation of natural resources and the environment. Preservation of the environmental heritage and natural resources for future generations requires the development of economically rational solutions which reduce the consumption of resources, halt environmental pollution and save natural ecosystems.

Social development. Actions should be focused on creating new jobs, ensuring access to food, education and medical care. This requires involvement of the entire international community, while taking care to preserve social and cultural diversity.

Moreover, the report indicates that sustainable development is possible at the current civilisational level, i.e. development in which the needs of the present generation can be met without reducing the chances of future generations to meet them (Brundtland 1989; Mondini 2019).

Another example of actions taken on a global political scale, indicating the need for transforming the existing policies, was the 1992 UN Conference held in Rio de Janeiro under the title "Environment and Development", which was attended by representatives of governments from 172 countries of the world and about 2400 representatives of non-governmental organisations (NGOs). During the debate known as the "Earth Summit", the idea of "sustainable development" was clarified. Since 1992, sustainable development has been a buzzword permanently present in international debates, and has become, next to environmental protection, a basic function of the state. Heal (1998) suggests that the essence of sustainable development is defined by the following three axioms:

- Such treatment of the present and future with, in the long term, has a positive value.
- Recognition of all the ways in which environmental assets contribute to economic welfare.
- Recognition of the limitations of environmental asset dynamics.

The 1992 UN initiative is considered to have launched the era of eco-development; as a result of lengthy negotiations and agreements, largely conducted before the Rio conference itself, the following documents were adopted by all participants: Rio Declaration on Environment and Development; Global Programme of Action—Agenda 21; Convention on Biological Diversity; Convention on Climate Protection; and Declaration on Forest Protection, known as the Forest Principles (Adede 1995; Momtaz 1996).

The issues addressed in the Declaration include: the right to the environment and to live in harmony, intergenerational fairness, international cooperation, environmental information and access to resources. Interestingly, these issues are present in the main economic doctrines that dominate contemporary politics. References to them can be found in liberal concepts as well as those referring to the need for state interventionism. This has resulted in, among other things, ecological taxes as well as state subsidies for organic farming. Although in the twentieth and twenty-first century one can observe an increasing clash between the extreme liberal approach of the Chicago school of economics and the followers of John Maynard Keynes's views (creator of the theory of state interventionism), this does not change the fact that the issues of green transition are also an important element of this discussion.

In the context of global political debates, another UN initiative should be mentioned. In 2000, the United Nations Millennium Summit defined 8 Millennium Development Goals that were to be achieved by 2015. Among them the following can be distinguished (Summit 2000):

1. Eradicate extreme poverty and hunger;
2. Achieve universal primary education;

3. Promote gender equality and empower women;
4. Reduce child mortality;
5. Improve maternal health;
6. Combat AIDS, malaria and other diseases;
7. Ensure environmental sustainability;
8. Develop a global partnership for development.

From the point of view of the international community, the Millennium Declaration, Millennium Development Goals (MDGs) and the international discussion of key targets have contributed immensely to raising public awareness, increasing political will and mobilising resources for energy transition.

An important document on transition and sustainable development was the 2030 Agenda. It was one of the outcomes of the UN Summit on Sustainable Development, which took place in September 2015 in New York. The 2030 Agenda is referred to as the world development strategy until 2030. It comprises 17 Sustainable Development Goals (GA 2015, p. 2030). It was adopted in 2015 when all 193 UN member states unanimously adopted the resolution entitled "Transforming our world: the 2030 Agenda for Sustainable Development" (Lee et al. 2016; Weiland et al. 2021). For each goal, specific tasks to be accomplished by 2030 were outlined—a total of 169 tasks. As a continuation of global efforts to improve the quality of life for all people in the world, the 2030 Agenda refers to and largely builds on the Millennium Development Goals pursued between 2000 and 2015. However, its scope is much broader than that of the Millennium Programme. Alongside priorities such as health, education, nutrition and food security, the Agenda identifies a number of economic, social and environmental goals, aiming to transform economies in such a way as to lay the foundations for long-term, sustainable, job-creating growth (Palmer 2015; Diaz-Sarachaga et al. 2018).

The exceptional consensus of the international community on prompt adoption of the 2030 Agenda was due to universality of its provisions. Ongoing discussion in this regard at the level of UN member states allowed defining the Agenda as a plan that would be flexibly adapted to the realities of each country. This also means imposing a specific role on states in terms of implementing its goals. The political discussions recognised that the complexity of global problems required broad commitment and responsible cooperation, so it was important to translate global goals into national, regional and local contexts and to raise awareness of their importance for each country.

Another milestone in the global discussion on the need to transform existing energy policies was the 2015 United Nations Climate Change Conference in Paris.

The aim of the conference was to establish an agreement and universal consensus on climate among all countries of the world and seek to intensify the global response to the threat of climate change, in the context of sustainable development and poverty eradication efforts, among other things, by (Robbins 2016; Dejuán et al. 2017; Hickmann et al. 2021):

- limiting increases in the global average temperature to much below 2 °C above pre-industrial levels and undertaking efforts to limit the temperature increase to 1.5 °C above pre-industrial levels, recognising that this will significantly reduce the risks and impacts of climate change;
- enhancing the capacity to adapt to the negative impacts of climate change and promoting climate change resilience as well as low greenhouse gas emission development in a way that does not threaten food production;
- ensuring coherence of financial flows with a pathway towards low GHG emissions and climate change resilient development.

The Paris Agreement was a step towards establishing a global goal to increase the adaptive capacity, strengthen resilience and reduce vulnerability to climate change. Economically developed countries committed to provide assistance to developing ones to finance climate change mitigation actions. The political dimension of the conference was to create a platform for cooperation in taking action to expand education, training, public awareness, public participation and universal access to climate change information. The agreement entered into force on 4 November 2016, with 193 countries as parties.

While analysing what is currently happening in the world, one should emphasise that the climate and the environment are very hot topics today, and the need for energy transition is not always understood in the same way by everyone. A particular example of this was the 24th session of the Conference of the Parties to the United Nations Framework Convention on Climate Change (UNFCCC) (COP24), commonly referred to as the Climate Summit. It was held from 3 to 14 December 2018 in Katowice under the motto Changing together.

For 11 days, representatives of 196 countries and the European Union debated to reach an agreement on implementation of the goals set three years earlier in the Paris Agreement. The Climate Summit began with the Leaders' Summit, during which the assumptions of the "Just Transition" concept were presented, devoted to ensuring a fair and solidarity based transformation of the economic model that would protect the climate while maintaining economic security of states, economic development and valuable jobs. A package of solutions included in the "Driving Change Together Partnership" was also discussed, relating to the proposed partnership for actions towards the development of electromobility and zero-emission transport.

The outcomes of the climate summit are assessed in different ways. Politicians point to the success of the negotiations, stressing that the summit was an opportunity to discuss detailed and technical issues which made it possible to adopt the so-called "Katowice Package". More pessimistic observers pointed to a number of divergences in the positions of individual states as regards the directions of green transition, while welcoming the adoption of a specific book of obligations, i.e. a set of principles that would allow the climate agreement to function in practice. These principles are supposed to enable the monitoring of reporting by individual states on the measures they are taking to protect the climate.

4 The Need for Energy Transition in the Social Debate

One of the milestones on the road towards energy transition was the now somewhat forgotten Club of Rome's Report "Limits to Growth". The role of this study was and still is enormous. When in 1972 a group of scientists from the Massachusetts Institute of Technology, namely Donell H. Meadows, Dennis L. Meadows, Jørgen Randers and William W. Behrens III developed several scenarios for the future of civilization using proprietary and innovative models for the time, the world seemed puzzled and concerned (Meadows et al. 1972). The question posed by the authors: Is unlimited growth possible on a limited planet? Intrigued, troubled and prompted the debate.

The main subject of the political debate of the 1970s became the main assumptions of that report concerning the pace and directions of humanity's resource consumption and the inevitable, as it seemed, reaching the end of capacity to use those resources (Giddens 2009). While the awareness of existence of non-renewable resources could still be accepted by the societies of that time, the vision of inevitable catastrophe was overlooked by many. The authors of the report shocked the public by proving with the models and analyses presented that if the growth trends of world population, industrialisation, pollution, food production and resource consumption were to continue, the limits of this planet's growth would be reached within the following hundred years. Then, the most likely outcome would be a rather rapid and uncontrolled decline in both population and industrial production (Cline 1992; Perman et al. 2003; Pieterse 2010).

That publication is important not only because of the analyses and conclusions it presents but, above all, because in subsequent years it was supplemented with new data and hypotheses and became the basis for subsequent reports by the Club of Rome. It also sparked a lively debate lasting many years on limited natural resources or the risk of overpopulation. It stirred up numerous controversies and attracted criticism upon the researchers and was the subject of many polemical speeches. The most important thing, however, is that, like a number of earlier scientific papers and political statements, the "Limits to Growth" report has not been forgotten. The discussion of its theses continues to this day and, it seems, is the axis of the political debate on the energy transition (Marcuse 1998; Bonviu 2014; Camilleri 2020).

When, almost 50 years later, Gaya Herrington looked at the famous report, the discussion flared up again (Herrington 2021). According to Herrington, it is not possible to speak of the end of civilisation, but it is certain that if the current economic, social and energy solutions are maintained, the world will face deterioration of living standards and decline in food production. A solution to this situation seems to be the concept of degrowth, which rejects the need for perpetual growth. This issue was discussed during the World Economic Forum in 2021. The main postulate within the degrowth movement is the reduction of production and consumption (Kallis et al. 2012; Sekulova et al. 2013; Spangenberg 2014). Analysing the relationships between economic growth, environmental destruction and issues related to social inequalities, representatives of the movement argue that the development model of the global economy based on unlimited economic growth leads to environmental degradation and creates huge inequalities between the rich and the poor—both on the level of countries and within individual societies.

The topic of sustainable growth and energy transition has been recurring for several years as a key issue in the economic and political discussion among the participants of the WEF World Economic Forum. The year 2020 was particularly notable in this debate, when the young Swedish climate activist Greta Thunberg in her speech in Davos, addressing the world's political and economic elites, pointed out to them that "virtually nothing has been done for the climate". This voice concerned not only the years of neglect in this area but also, above all, referred to the lack of reflection and action related to the global Youth Climate Strike (Fridays for Future, Youth for Climate, Climate Strike, Youth Strike for Climate), which was inspired by Greta Thunberg's strikes in Sweden in 2018. In 2019, this protest by schoolchildren and students against and in the face of politicians' passivity towards global warming and man-made climate change spread to more than 150 countries. On 11 December 2019, Greta Thunberg was awarded the prestigious title of the Time magazine's Person of the Year. The activist recalled that many months had passed since the protests and the youth's demands had not been addressed in any way. It is now possible to speak explicitly about the GT Effect, which is defined as increased interest in climate issues among young people around the world (Sabherwal et al. 2021; Zhanda et al. 2021).

Therefore, it seems that the political and social activist Naomi Klein is right when she says that the key to accomplishment of energy transition is to evoke motivations other than those of climate concern. It is necessary to present energy transition not as a necessary sacrifice, but as an opportunity for companies, municipalities to make a better living, earn or save money. According to Klein, in the climate struggle it is clear that we will not gather sufficient forces to win if we do not base the climate policy on justice—first and foremost racial justice, but also gender and economic justice.

In her most recent speeches and publications, Klein emphasises that energy transition can entail social transformation. It offers a unique opportunity to transform the society into one that is more equitable in all respects and in which no person's life counts less than other people's lives (Klein 2020; Klein et al. 2020).

A similar view is taken by the Green Growth Knowledge Platform (GGKP). It is a global network of international organisations and experts that identifies major knowledge gaps in the theory and practice of green growth. According to the GGKP, green growth means combining the promotion of economic development with ensuring that the environment will provide the resources and services necessary for the future prosperity of mankind. To achieve this goal, it is necessary to develop an informed society that interacts with the market through certain attitudes and behaviours.

5 EU's Position on Energy Transition

The origins of the European Union's energy policy can be traced back to the founding Treaties of 1951 and 1957. Establishment of the European Coal and Steel Community, followed by the European Economic Community and the European Atomic Energy Community, had not only an economic but also a political dimension. However, it was not until the first oil crisis in 1973 that prompted the Member States to begin the process of creating a common energy policy. This also meant that, from the outset, European countries were involved in the debate on necessary changes in the global economy and politics, aimed at reducing the use of natural resources. This is because it was in Europe that successive industrial revolutions began whose common denominator was maintaining constant access to sources of cheap energy (Capello and Cerisola 2022; O'Brien 2022). Basing the European economy on coal, oil and nuclear energy was for years a guarantee of civilisational progress, high quality of life and high competitiveness of European economies (Laitner 2000; Voigtländer and Voth 2006). Thanks to access to cheap energy, European countries achieved a technological and qualitative advantage needed to develop mass production. In this respect, they were ahead of other countries that had more limited access to cheap energy and high technologies. At the same time, there was a clear process of European economies becoming more dependent on external supplies of increasingly expensive oil and natural gas. The depletion of indigenous resources and rising energy consumption necessitated strategic decisions regarding the EU's energy transition. However, this process was not easy. From the very beginning, diverging interests of the Member States were a significant obstacle to achieving common goals. This was particularly true of the positions of the French and British governments, which for many years prioritised bilateral agreements and solutions over the need to create coherent solutions within the EU.

The spectacular shift towards a green European economy is based on more efficient energy use and progressive replacement of fossil fuels with clean energy sources. This process is understood as a shift from the current energy system using non-renewable/traditional energy sources (fossil fuels) to an energy system based mainly on renewable and zero-/low-carbon sources. This is to allow the European economy to become independent from traditional energy sources i.e. oil and natural gas. This is particularly important from the economic and strategic point of view, and for ensuring Europe's energy security. The EU's dependence on energy imports, especially for oil and natural gas, is 55%, but by modernising the energy generation sector, it is estimated to fall to 20% by 2050 (Petersen et al. 2021; Keppo et al. 2022). This will significantly reduce the expenditure on energy resources imports, while at the same time ensuring a higher level of energy self-sufficiency. Successive crises on the global fuel market, erupting conflicts in areas from which cheap oil and gas have traditionally been sourced, and the desire to become independent from Russian sources of supply have all brought about changes in the European energy policy. European countries have come a long way from the European Coal and Steel Community and the Council for Mutual Economic Assistance, whose functioning was based on the development of coal and lignite mining, to the European Union preaching accelerated decarbonisation of the economy (Oberthür and Dupont 2015; Drummond and Ekins 2017; Golombek et al. 2022).

The current energy transition model is understood as conversion to a sustainable economy. Its assumptions are based on achieving high energy efficiency of low-carbon energy sources and electromobility. Consistent implementation of this model in Europe, forcing similar changes in other countries around the world, will allow not only to maintain high competitiveness of European economies, but also to achieve the expected positive environmental and socio-political effects. Energy transition in the European dimension is based on the desire to increase energy security. The role of this kind of security has particularly increased in the twenty-first century. Destabilisation in the so-called Middle East, the collapse of Venezuela as a major oil producer, the conflict in Ukraine of 2014 and 2022 are just a few examples of what has significantly changed thinking about the future of European economies over a period of two decades.

However, the European energy transition goes beyond energy security and entails a number of new challenges, including inevitable changes for the society and the economy. Their economic dimension is based on new investment opportunities and potential exports of low-carbon technologies, transformation of industry and shift towards less carbon-intensive technologies, including greater digitisation of processes.

The solutions being introduced include gradual shift from exhaustible hydrocarbons and uranium fuel to RES. This applies to all sectors of the economy (transport, industry, energy, agriculture, etc.).

In the social dimension, the ongoing political discussion focuses on the necessity of climate protection measures and provision of public health benefits.

The energy transition in the EU was the result of the informed environmental policy implemented since the 1980s. Successive climate summits and reports made these problems a hot topic of the political and social debate in the unifying Europe. In preparation for new challenges, the countries of the European Communities adopted a key document for the energy transition, the Single European Act, in 1987. The Single European Act was the first major modification of the 1957 Treaty of Rome. The Act formally instituted the creation of a common European market but, above all, it strengthened political cooperation among the countries of the Community (Knill and Liefferink 2012). This resulted in new environmental legislation, which stipulated that the EU Council could decide jointly with the European Parliament on environmental protection legislation.

An important event, from the point of view of the current model of energy transition in Europe, was the negotiation by member states in 1995 of the "Green Paper on a common energy policy" (Leydon 1995; Maltby 2013). This energy consensus established Community principles for creation of the European energy market and related environmental protection issues. The Green Paper became a key document in the political debate on the model of energy transition in the EU and contributed to the growing interest in environmental and climate change issues. This was reflected, inter alia, in the political declaration of the EU countries on the implementation of the 1997 Kyoto Conference. Under the Kyoto Protocol, EU countries (i.e. EU15) committed to reduce greenhouse gas emissions by 8% on average between 2008 and 2012 compared to 1990 (Mlynarski 2017). Meanwhile, the other industrialised countries participating in the agreement declared a 5% reduction in emissions.

The political discussion resulted in further declarations and strategies, among which the following should be mentioned:

- "White Paper—Energy for the Future" of 1997;
- "Green Paper on the Security of Resources" of 2000;
- "European Security Strategy—a Secure Europe in a Better World" of 2003;
- "Green Paper—a European Strategy for Sustainable, Competitive and Secure Energy" of 2006.

The Green Paper adopted in 2006 was to be a major step towards creating the conditions for green energy development in the EU. In this document, the European Commission indicated 6 key areas of the energy policy for the EU:

- Finalising the creation of European internal markets for electricity and gas;
- Solidarity among member states;
- Security and competitiveness of energy supplies;
- Climate changes;
- Innovation in the energy technologies;
- Coherent external energy policy.

This was related not only to growing awareness of the need to implement environmentally friendly solutions in the European economy, but also to the consequences of another energy crisis in 2006, caused by reduced gas supplies from Ukraine.

In December 2008, the European Council adopted the climate and energy package, which set a CO_2 reduction target of 20% by 2020.

The ongoing process of political integration within the EU, manifested by the Lisbon Treaty of 2009, has had a significant impact on the EU's environmental protection policy. The Lisbon Treaty gave the EU exclusive competence over the environment, which had until then been shared between the Member States and the EU. The Treaty also indicated that the EU aimed to combat climate change and promote measures on the international level to deal with regional or global environmental problems.

The adopted solutions created a solid foundation of the Community's climate policy, which included the EU Emissions Trading Scheme (EU ETS). It has become the main mechanism for combating climate changes in the EU. The EU ETS is a key tool for reducing greenhouse gas emissions in a cost-effective manner. It is the first and, until 2017, the largest and currently the second (largest after China) market in the world for carbon allowances. At present, the EU ETS is the only CO_2 market based on binding national and regional legislation. Carbon trading is based on provisions in the Kyoto Protocol to the United Nations Framework Convention on Climate Change (UNFCCC) adopted in 1997 (Ellerman et al. 2014; Rosendahl 2019).

Another challenge was put before the EU Member States in 2014. At the summit of the European Council, heads of state set new climate and energy targets until

2030. More ambitious reduction targets, including a 40% reduction in greenhouse gas emissions compared to 1990, were the subject of particular political debate, as was the commitment to source at least 27% of energy from renewable sources and to improve energy efficiency. These assumptions were proposed during the UN Framework Convention on Climate Change (UNFCCC). They resulted in the EU's accession to a new global climate agreement (UNFCCC Paris Protocol).

The Paris Agreement laid the groundwork for the European Green Deal policy adopted by the European Commission on 11 December 2019, with the overarching goal of achieving climate neutrality for Europe by 2050 (Kemfert 2019; Maris and Flouros 2021). Discussions in the political and social space accompanying implementation of the Paris Agreements increasingly emphasised that climate change and environmental degradation posed a threat to Europe and the rest of the world. To meet these challenges, the European Green Deal Action Plan was created. It is intended to help transform the EU into a modern, resource-efficient and competitive economy with net zero greenhouse gas emissions as early as 2050. The premise of the new economic model was to move away from growth based on resources consumption and to ensure even greater solidarity between people and regions (Austvik 2016).

As part of implementation of the Green Deal, the European Commission adopted the "Fit for 55" package on 14 July 2021, consisting of 13 legislative proposals to align EU policies to reducing greenhouse gas emissions. These include:

- revision of the EU's CO_2 emissions trading mechanism (EU ETS);
- introduction of a new mechanism for trading CO_2 emissions rights for buildings and transport—so-called mini-ETS);
- introduction of a cross-border adjustment mechanism (CBAM) at the EU border to ensure competitiveness of European producers of energy-intensive goods vis-à-vis operators from countries where emission rights are not paid for;
- measures relating to the Effort Sharing Regulation, binding targets for Member States on GHG emissions from sectors not covered by the EU ETS system;
- amendment to the Energy Taxation Directive, which sets out the institutional and legal framework for taxation of electricity and energy products;
- amendment to the RES Directive (RED II), which promotes the use of energy from renewable sources. The aim is to increase RES from 32 to 40% of final energy consumption by 2030;
- amendment to the Energy Efficiency Directive (EED), promoting measures to increase energy efficiency;
- amendment to the regulation setting CO_2 emission standards for new cars and vans.

One of the tools for implementing the principles of the Green Deal are to be actions of the European Commission aimed at encouraging European manufacturers to switch to a circular economy. In particular, this concerns industries that consume significant quantities of raw materials, such as the clothing, construction, electronics and plastics production industries. Among other things, the European Commission wants all packaging used in the Community to be recyclable by 2030. Moreover, it plans to place emphasis on the production of reusable, durable and repairable products.

According to the EC's calculations, transport is responsible for 25% of greenhouse gas emissions in the EU. Therefore, in order to reduce CO_2 emissions into the atmosphere to zero, it is necessary to reduce emissions from vehicles, ships and so on by 90%. This is a huge challenge in the field of transport innovation, including electromobility and hydrogen cells. According to the authors of the Green Deal, the European Union is able to achieve the goals set, but this requires very high public expenditure and actions that will convince entrepreneurs to invest in environmental protection.

Being aware of the necessity to incur high expenditures in order to transform the economy towards a climate neutral one, the decision was taken to allocate at least EUR 1 trillion over the next decade to implement the necessary measures. The European Commission proposed that 25% of the EU budget for 2021–2027 should be allocated to climate action and environmental spending across a range of programmes.

Implementation of the "Fit for 55" package is the greatest challenge for the European economy. Estimated annual investment costs in the energy sector to reduce CO_2 emissions by 55% are EUR 350 billion higher than in 2011–2020, with total investment of almost EUR 5 trillion by 2030. This capital boost will help rebuild the economy after the current COVID-19 crisis and accelerate the transition towards a climate neutral economy. The EU is scheduled to become the first climate neutral territory by 2050.

Climate neutrality means a balance between greenhouse gas emissions and their absorption from the air. Reducing emissions on the one hand and using carbon sinks on the other is therefore crucial. Green transition represents a huge opportunity for the European industry by creating new markets for new green technologies and products. Electrification of the economy and massive use of renewable energy sources are expected to boost employment in many sectors of the European economy. The Green Deal is thus the most ambitious and largest project of European integration in its history. Introduction of the Green Deal will certainly trigger many new discussions, conflicts and tensions. It seems, however, that striving for climate neutrality will be an undisputed priority in the coming years.

6 EU's Energy Security in the Light of the Conflict in Ukraine

The aggression of the Russian Federation against Ukraine has highlighted the dependence of European economies on the supply of energy resources from Russia. The outbreak of war in Ukraine in 2022 means that EU countries have had to review their energy policies quickly. It seems that many of them are now making key decisions on implementing climate neutrality and other socio-economic policies. Solidarity in imposing political and economic sanctions on Russia indicates that, in the energy sphere too, one should expect significant changes in the policies implemented at both the European and national levels. The need to increase energy security has become a priority topic in political discussions in the countries of the united Europe.

Actions aimed at changing the energy systems of EU member states depend on the structure of the energy balance, including the role of high-emission fossil fuels (coal, natural gas, oil) in the economies of individual countries. In the context of energy security, there also arises the problem of high dependence of individual economies on gas and oil imports from Russia.

According to the green transition strategy, the target fuel to ensure climate neutrality was to be hydrogen. However, until new, efficient technologies using hydrogen were developed, gas was to be the transition fuel to which some European countries switched their energy production. This was reflected in reduced coal use in electricity generation and increased use of gas—see Fig. 2.

In 2021, the European Union imported 155 bln m^3 of gas from Russia, which accounted for about 45% of EU gas imports in 2021 and almost 40% of total gas consumption.

The dilemma currently facing European countries is not how high gas prices are, but whether it will flow to consumers at all. In particular, this concerns the economies of Greece, Austria, Hungary, Italy, the Netherlands, Germany and Moldova, the latter being completely dependent on gas supplies—see Fig. 3. These countries rely heavily on natural gas for their electricity production, making them highly vulnerable to the Russian energy blackmail.

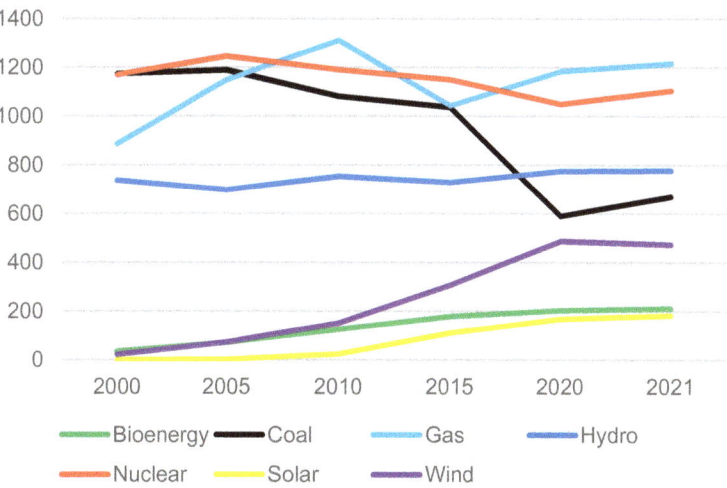

Fig. 2 Europe electricity generation by source [in Terawatt hours]. *Source* Own compilation based on data from Ember https://ember-climate.org

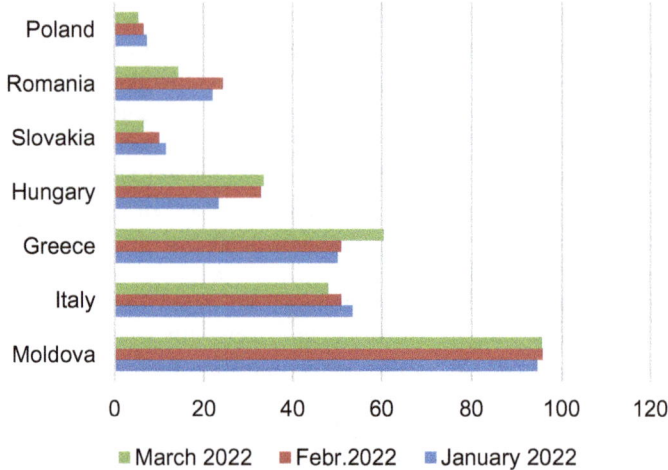

Fig. 3 Share of gas based energy generation in total electricity generation in the first three months of 2022. *Source* Own compilation based on data from ENTSO-E

According to European Commission's estimates, Russia has captured 40% of the gas market in the Union. According to Eurostat data, in 2021, gas from Russia will account for as much as 93% of gas imports.

One of the solutions under discussion is the creation of a common purchasing platform as soon as in May 2022, where the EU will negotiate gas supply contracts for member states plus Ukraine, Georgia and Moldova. This is a return to the idea of an Energy Union in Europe, already put forward in 2010 by the President of the European Parliament Jerzy Buzek (see: Austvik 2016).

The consequence of the conflict in Ukraine is the statement that the EU wants to become independent from Russian gas by 2027. This is particularly important for those countries where gas imports from Russia account for a significant share of domestic consumption—see Fig. 4.

Some member states are declaring an earlier date for giving up gas imports from Russia. Poland assumes that this will happen already in 2022. In March 2022, Robert Habeck, the German Vice-Chancellor and Minister for the Economy, indicated that the German economy would become independent from Russian coal and oil imports as soon as in 2022.

Threats concerning the provision of supplies mean that the current political debate focuses on swift start-up of the Baltic-Pipe pipeline and expansion of LNG terminals. This is supposed to make the continent independent from one supplier, which has proven to be unpredictable and dangerous in political and economic terms for the whole world.

The United States is to play a key role on the road to independence from Russia. President Joe Biden signed an agreement with the European Commission to increase LNG supplies to European terminals. LNG imports from the USA are to reach 50 bln

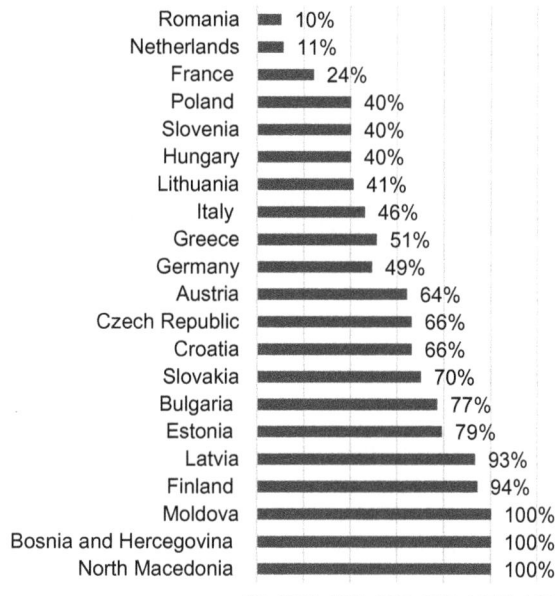

Fig. 4 Gas imports from Russia as % of domestic consumption. *Source* Own compilation based on data from ENTSO-E

m^3, up from 2 bln m^3 in 2021. Negotiations are underway to increase LNG imports from Qatar, Azerbaijan, Nigeria or Egypt.

The changing geopolitical situation also means that EU member states are forced to base their economies to a greater extent on an energy mix that includes, in addition to RES, nuclear energy and, temporarily, hard coal and lignite. However, it is possible to significantly reduce imports of energy resources from Russia, which may become a tool of political pressure on the Kremlin. This applies in particular to gas. According to the International Energy Agency (IEA), this is possible thanks to introduction of a 10-point programme which is to allow the European Union to become independent from Russian gas supplies. The project's assumptions are fully consistent with the European Green Deal, Fit for 55 package and the EU's climate ambitions contained therein. Key actions recommended by the IEA include:

- no new gas supply contracts with Russia;
- replace Russian supplies with gas from alternative sources;
- introduce minimum gas storage obligations;
- accelerate the deployment of new wind and solar projects;
- maximise generation from existing dispatchable low-emission sources: bioenergy and nuclear;
- enact short-term tax measures on extraordinary gains to shelter vulnerable electricity consumers from high prices;
- speed up the replacement of gas boilers with heat pumps;
- accelerate energy efficiency improvements in the building industry;

- encourage a temporary 1 °C thermostat adjustment by consumers;
- step up efforts to diversify and decarbonise sources of power system flexibility.

According to the IEA, these measures could reduce Russian gas imports to the EU by more than 50 bln m^3 per year.

In the ongoing debate, strategic decisions on rapid independence from imports of energy resources from Russia are also encountering opposition from some countries that have strong political and economic ties with Russia. This particularly concerns Hungary and Slovakia.

7 Summary

The discussion on the need for green transition of the global economy, which has been going on for years, has shown that in order to achieve the objectives set, many different options and actions need to be analysed. In particular, this concerns their effectiveness, potential impact on the quality of life, social and economic sustainability. As with previous energy transitions, this is now an extremely difficult and complex issue.

The political debate on creating a model for the use of non-renewable resources is conducted on the forum of global integration organisations, as well as in the social dimension. It involves both the academic community and non-governmental organisations. The common goal is to implement the principles of sustainable development on the global level and, thus, to achieve transition not only in terms of energy sources, but also in the broad context of the global economy.

Of particular importance in this process are activities aimed at providing a fast and effective growth impulse and creating new opportunities for the world economy, struggling with the Covid19 crisis or armed conflicts taking place in different parts of the world.

In summary, one can say that they are mainly concerned with diversity and technological imbalances. This leads the highly developed countries to try to impose certain solutions. In this case, it is important to create a dynamic and flexible energy system, based on diverse technologies that meet the needs of different groups of countries.

Analysis of the green transition measures already taken prompts reflection on the need for systemic solutions to manage a diversified system that is the result of a mix of technologies and nationally differentiated economic potential. New ways must be found to balance supply and demand in a diversified energy system that includes traditional energy carriers such as gas, coal and oil, as well as renewable sources such as solar and wind energy.

Creating effectively enforced regulations is also crucial to success of the measures taken. There is a need for a strong carbon pricing mechanism through a combination of regulatory measures and incentives. This should encourage societies to step up their green transition efforts. In addition, the revenue generated in this way should support equitable transition that takes account of social and economic circumstances. It is also important to support innovation leading to the decoupling of non-renewable energy sources and leading to implementation of a zero emission economy model.

References

Adede AO (1995) The treaty system from Stockhold (1972) to Rio de Janeiro (1992). Pace Envtl L Rev 13:33
Austvik OG (2016) The Energy Union and security-of-gas supply. Energy Policy 96:372–382
Barkemeyer R et al (2014) What happened to the 'development' in sustainable development? Business guidelines two decades after Brundtland. Sustain Dev 22(1):15–32
Biermann F, Kanie N, Kim RE (2017) Global governance by goal-setting: the novel approach of the UN Sustainable Development Goals. Curr Opin Environ Sustain 26:26–31
Bonviu F (2014) The European economy: from a linear to a circular economy. Rom J Eur Aff 14:78
Brundtland GH (1989) Global change and our common future. Environ: Sci Policy Sustain Dev 31(5):16–43
Buchanan JM, Buchanan JM Tullock G (1965) The calculus of consent: logical foundations of constitutional democracy. University of Michigan Press
Camilleri MA (2020) European environment policy for the circular economy: implications for business and industry stakeholders. Sustain Dev 28(6):1804–1812
Capello R, Cerisola S (2022) Industrial transformations and regional inequalities in Europe. Ann Reg Sci 1–14
Cline WR (1992) Economics of global warming, the. Peterson Institute Press: All Books [Preprint]
Conca K (1995) Greening the United Nations: environmental organisations and the UN system. Third World Q 16(3):441–458
Coyle ED, Simmons RA (2014) Understanding the global energy crisis. Purdue University Press
Dejuán Ó, Lenzen M, Cadarso MÁ (2017) Environmental and economic impacts of decarbonization: input-output studies on the consequences of the 2015 Paris Agreements. Routledge
Diaz-Sarachaga JM, Jato-Espino D, Castro-Fresno D (2018) Is the Sustainable Development Goals (SDG) index an adequate to measure the progress of the 2030 Agenda? Sustain Dev 26(6):663–671
Dooley MP, Folkerts-Landau D, Garber P (2004) The revived Bretton Woods system. Int J Financ Econ 9(4):307–313
Drummond P, Ekins P (2017) Cost-effective decarbonization in the EU: an overview of policy suitability. Clim Policy 17(sup1):S51–S71
Ellerman AD, Marcantonini C, Zaklan A (2014) The EU ETS: Eight years and counting. Robert Schuman Centre for Advanced Studies Research Paper [Preprint], (2014/04)
Falkner R (2012) Global environmentalism and the greening of international society. Int Aff 88(3):503–522.
GA U (2015) Transforming our world: the 2030 Agenda for Sustainable Development. Division for Sustainable Development Goals, New York, NY, USA [Preprint]
Giddens A (2009) Politics of climate change. Polity
Golombek R et al (2022) The role of transmission and energy storage in European decarbonization towards 2050. Energy 239:122159
Gomółka M (2010) Przyszłość koncepcji międzynarodowej walki z globalnym ociepleniem w kontekście debaty o przyczynach zmian klimatu. Kultura i Polityka 7:121–133
Hall SG, Tavlas GS (2013) The debate about the revived Bretton-Woods regime: a survey and extension of the literature. J Econ Surv 27(2):340–363
Handl G (2012) Declaration of the United Nations conference on the human environment (Stockholm Declaration), 1972 and the Rio Declaration on Environment and Development, 1992. United Nations Audiovisual Library of International Law, 11, p 6

Heal GM (1998) Valuing the future: economic theory and sustainability. Columbia University Press
Herrington G (2021) Update to limits to growth: comparing the World3 model with empirical data. J Ind Ecol 25(3):614–626
Hickmann T et al (2021) The United Nations Framework Convention on Climate Change Secretariat as an orchestrator in global climate policymaking. Int Rev Adm Sci 87(1):21–38
Hultman M, Pulé PM (2018) Ecological masculinities: theoretical foundations and practical guidance. Routledge
Igwe IO (2018) History of the international economy: the Bretton Woods system and its impact on the economic development of developing countries. Athens J Law 4:105
Kallis G, Kerschner C, Martinez-Alier J (2012) The economics of degrowth. In: Ecological economics. Elsevier
Kasperson RE, Dow KM (1991) Developmental and geographical equity in global environmental change: a framework for analysis. Eval Rev 15(1):149–170
Katz-Rosene R, Paterson M (2018) Thinking ecologically about the global political economy. Routledge
Kemfert C (2019) Green deal for Europe: more climate protection and fewer fossil fuel wars. Intereconomics 54(6):353–358
Keppo I et al (2022) Introduction to EMP-E 2019 special issue modelling the implementation of 'a clean planet for all' strategy. Energy Strategy Rev [Preprint]
Klein N, Ramos TB, Deutz P (2020) Circular economy practices and strategies in public sector organizations: an integrative review. Sustainability 12(10):4181
Klein N (2020) On fire: The (burning) case for a green new deal. Simon & Schuster
Knill C, Liefferink D (2012) The establishment of EU environmental policy. In: Environmental policy in the EU. Routledge, pp 39–57
Koc TC, Teker S (2019) Industrial revolutions and its effects on quality of life. PressAcademia Procedia 9(1):304–311
Kozyra P, Rutkowska M, Rembielak-Vitchev G (2007) Evolution of European Union ecological policy–in environmental programs actions
Kreibich R (2007) All tomorrow's crises. IP—Global Edition, pp 11–15
Laitner J (2000) Structural change and economic growth. Rev Econ Stud 67(3):545–561
Lee BX et al (2016) Transforming our world: implementing the 2030 agenda through sustainable development goal indicators. J Public Health Policy 37(1):13–31
Leydon K (1995) For a European Union energy policy: the Energy Green Paper: address to the Parliamentary Group for Energy Studies. Energy Focus 12
Maltby T (2013) European Union energy policy integration: a case of European Commission policy entrepreneurship and increasing supranationalism. Energy Policy 55:435–444
Marcuse P (1998) Sustainability is not enough. Environ Urban 10(2):103–112
Maris G, Flouros F (2021) The green deal, national energy and climate plans in Europe: Member States' compliance and strategies. Admin Sci 11(3):75
Meadows DH et al (1972) The limits to growth: a report to the club of Rome (1972). Google Scholar 91
Mlynarski T (2017) Bezpieczeństwo energetyczne i ochrona klimatu w drugiej dekadzie XXI wieku: energia-środowisko-klimat. Wydawnictwo Uniwersytetu Jagiellońskiego, Kraków
Momtaz D (1996) The United Nations and the protection of the environment: from Stockholm to Rio de Janeiro. Polit Geogr 15(3–4):261–271
Mondini G (2019) Sustainability assessment: from brundtland report to sustainable development goals. Valori e Valutazioni [Preprint], (23)
Nordhaus WD (1980) The energy crisis and macroeconomic policy. Energy J 1(1)
Nurzyńska A (2017) ONZ–order and safety world. World Sci News 78:213–219
O'Brien P (2022) Was the British industrial revolution a conjuncture in global economic history? J Glob Hist 17(1):128–150
Oberthür S, Dupont C (2015) Decarbonization in the European Union: internal policies and external strategies. Springer

Palmer E (2015) Introduction: the 2030 agenda. J Glob Ethics 11(3):262–269
Pasten C, Santamarina JC (2012) Energy and quality of life. Energy Policy 49:468–476
Pearce D, Atkinson G (1998) The concept of sustainable development: an evaluation of its usefulness ten years after Brundtland. Rev Suisse Economie Politique Stat 134:251–270
Perman R et al (2003) Natural resource and environmental economics. Pearson Education
Petersen UR et al (2021) Documentation-The European Commission's "A clean planet for all" scenarios modelled in EnergyPLAN
Pieterse JN (2010) Development theory. Sage
Robbins A (2016) How to understand the results of the climate change summit: conference of Parties21 (COP21) Paris 2015. J Public Health Policy. Springer
Rosendahl KE (2019) EU ETS and the waterbed effect. Nat Clim Chang 9(10):734–735
Sabherwal A et al (2021) The Greta Thunberg Effect: familiarity with Greta Thunberg predicts intentions to engage in climate activism in the United States. J Appl Soc Psychol 51(4):321–333
Sand PH (2007) The evolution of international environmental law. In: The Oxford handbook of international environmental law
Sekulova F et al (2013) Degrowth: from theory to practice. J Clean Prod 38:1–6
Singer SF (2008) Nature, not human activity, rules the climate
Singer SF, Idso C (2009) Climate change reconsidered: the report of the Nongovernmental International Panel on Climate Change (NIPCC). The Heartland Institute, Chicago [Preprint]
Sneddon C, Howarth RB, Norgaard RB (2006) Sustainable development in a post-Brundtland world. Ecol Econ 57(2):253–268
Sohn LB (1973) The Stockholm declaration on the human environment. Harv Int'l LJ 14:423
Spangenberg JH (2014) Institutional change for strong sustainable consumption: sustainable consumption and the degrowth economy. Sustain: Sci, Pract Policy 10(1):62–77
Summit M (2000) United Nations Millennium Declaration. In: U. Nations (ed) [Preprint]
Thant SU (1971) The United Nations and some problems of public understanding. United Nations Office of Public Information
Voigtländer N, Voth H-J (2006) Why England? Demographic factors, structural change and physical capital accumulation during the industrial revolution. J Econ Growth 11(4):319–361
Weiland S et al (2021) The 2030 Agenda for Sustainable Development: transformative change through the sustainable development goals? Polit Gov 9(1):90–95
Young OR et al (2008) Institutions and environmental change: principal findings, applications, and research frontiers. MIT Press, Cambridge, MA
Zhanda K, Dzvimbo MA, Chitongo L (2021) Children climate change activism and protests in Africa: reflections and lessons from Greta Thunberg. Bull Sci Technol Soc 41(4):87–98

The Role of Green Energy in the Economic Growth of the World

Irena Łącka

Abstract The demand for energy accompanying economic growth is increasing. With natural resources depleting at an accelerating rate and, at the same time, the costs of exploiting them rising, it is clear that the search is on for energy sources that respect the economic aspects of energy production and also focus on environmental protection. The main purpose of using green energy in the economy is to minimise the negative human impact on the environment. After Russia's aggression against Ukraine in February 2022, the need for resigning from various energy resources (oil, gas, coal) from Russia and provision of the energy security of European countries based on renewable energies and green hydrogen became the factor accelerating the transition in Europe. The progress in transforming the traditional economy towards "the green direction" is evident in highly developed countries; by developing green energies, they become the direct beneficiaries of advantages related to their implementation. The chapter describes the reasons, manners of energy transition of the economy and its conditions. Furthermore, it describes the most important advantages and barriers resulting from the green transformation of the economy.

1 Introduction

For years, together with the increase in the global population and the increased developmental level of particular countries (previously less developed ones and consuming less energy), the demand for energy in the world has been increasingly growing. According to the International Energy Agency (2021) data, in the years 2020–2050, the world energy consumption will increase by 50%, while the global carbon emissions (CO_2) will increase by 25% in this period. Adjusting energy supplies to their increasing demand is a global problem and a significant developmental factor for many countries. It results from the fact that the energy supply is mainly satisfied by a continuous increase in the use of fossil fuels (gas, coal and crude oil), the resources of which are limited and finite in the long-term perspective. Their combustion releases

---

I. Łącka (✉)
Faculty of Economics, West Pomeranian University of Technology in Szczecin, Szczecin, Poland
e-mail: irena.lacka@zut.edu.pl

© The Author(s), under exclusive license to Springer Nature Switzerland AG 2023
I. Bąk and K. Cheba (eds.), *Green Energy*, Green Energy and Technology,
https://doi.org/10.1007/978-3-031-12531-7_3

thermal energy, which is next transformed into electrical energy. This process emits large amounts of carbon dioxide into the atmosphere (Moorthy et al. 2019). It is one of the greenhouse gases that contribute to global warming and, consequently, to climate changes, dramatic deterioration of the natural environment, and the severe limitation of opportunity for economic and social development of humanity in the future.

For years, more and more often, societies of many countries, including highly developed ones, have been experiencing the consequences of growing crises—ecological and energy crises, consequences of global warming in the form of extreme climatic phenomena (including tornadoes and hurricanes, fires, droughts, heavy rains, huge floods, very high temperatures). Inhabitants of these countries perceive growing threats to ecosystems and species, deterioration of living conditions and food production in successive, increasingly larger areas (Carayannis et al. 2012; Hoegh-Guldberg et al. 2018). As a result, millions of people are migrating in search of new, safer places to live, and ethnic or interstate conflicts over resources and territories are escalating.

At the same time, the changes in the world economy indicate that "Over the next decade, every aspect of national energy systems will be affected by changes in climate and energy policy, and financing, continuous technological advancement, and shifts in energy supply and demand. The rapidly falling costs of renewable technologies have opened up previously unimagined possibilities across the globe. Ongoing developments in many countries offer a promising outlook for the security, inclusiveness, and sustainability inherent in a transformed energy sector" (Theme Report on Energy … 2021). The above-mentioned changes resulted from the choice made at the end of the early 1900s of "green economy" and "green growth" as a way out of the economic crisis (Bąk and Cheba 2020; Tănasie et al. 2022).

The contribution of these various factors and the need for economic recovery after the COVID-19 pandemic (Łącka and Wojdyła 2022) changed the attitudes of governments of various countries and their societies toward the sustainable development. In most cases, in highly developed countries, the acceptance of the change of development paradigm and looking for a new economic growth model increased (Dacko et al. 2020: 135–162). It is directed towards sustainable development with a simultaneous stabilisation of material goods consumption and providing future generations with the possibility of satisfying their needs (Łuniewski and Łuniewski 2020). One element of building such an economy is the energy transition and decarbonisation of every sector—from industry to energy to transportation.

The reconstruction of the economy towards "green"—sustainable—one is advanced to the maximum extent possible in the European Union. However, not all member states implement actions to achieve that goal simultaneously. Energy transition, decarbonisation and zero greenhouse gas neutrality were written down in strategic documents of the European Union, and in particular in the European Green Deal. They aim to achieve climate neutrality in 2050 (EC 2019; Henderson and Sen 2021; Mrozowska et al. 2021). In recent years, many UN and EU representatives have emphasised the need to accelerate the energy transition.

Initially, this new idea was treated as something far in the future. In the long-term planning horizon, it was supposed to lead to increased prosperity, improved quality of life and social equality while at the same time halting the depletion of natural resources and reducing ecological risks. However, due to the pace and manner of changes in the world's natural resources and the increasing phenomenon of global warming, the implementation of the new concept was accelerated. A significant decrease in ecological risk and transformations of the contemporary economy (its values and foundations) must occur faster.

The sustainable economy model, called the "green" economy, was presented in detail a dozen years ago within the framework of the United Nations Organization, and numerous international organisations, including the European Union strategy (e.g. UNEP 2011; ICC 2012; EC 2010). The European Union became a leader in implementing the green economy principles, reflected in the Union's long-term development strategy entitled the European Green Deal (EC 2019), prepared in 2019 and the so-called hydrogen strategy. The latter indicated that the production and use of the "green" hydrogen are supposed to be a leading technology of the green transformation in Europe (EC 2020). In this situation, the green economy ceased to be only an idealistic concept. If it was to make the intended goals possible in the long planning horizon (to bring about an increase in prosperity, improve the quality of life and social equality, halt the depletion of natural resources and reduce ecological risks), then faster changes were needed in various spheres of the economy's functioning.

2 The Concept and Evaluation of Energy Transition in the World

One of the elements of such reconstruction of the world economy and national economies is the energy transformation. It is a process of significant structural changes in the energy system (Henderson and Sen 2021), giving rise to numerous challenges in many dimensions: technological, economic, social, financial, institutional and regulatory. According to the definition of the International Renewable Energy Agency (IRENA), "The energy transition is a pathway toward the transformation of the global energy sector from fossil-based to zero-carbon by the second half of this century. At its heart is the need to reduce energy-related CO_2 emissions to limit climate change" (Energy transition, https://www.irena.org/energytransition).

Decarbonisation and the shifting to a low carbon economy require shifting from energy sources based on hydrocarbons (crude oil, natural gas, hard coal and lignite) to renewable energy sources (Jing 2016). It is also necessary to save energy and improve energy efficiency. Energy transition should ultimately lead to replacing the energy from fossil, non-renewable natural resources with wind and biomass energy (e.g. biogas and landfill gas), water force and tides, solar (thermal and photovoltaic) energy as well as geothermal energy (Johansson et al. 2006; Turkenburg 2015: 221). We refer to this kind of energy as green energy. For some time, green hydrogen energy has also been indicated as an energy transition element (Wolf and Zander 2021), which has been confirmed by the hydrogen strategy adopted by the European Commission in

2020. "A Hydrogen strategy for a climate-neutral Europe" (EC 2020). As the World Economic Forum's experts indicate, „Green hydrogen could be a critical enabler of the global transition to sustainable energy and net-zero emissions economies. There is unprecedented momentum worldwide to fulfil hydrogen's longstanding potential as a clean energy solution. [...] Hydrogen is emerging as one of the leading options for storing energy from renewables with hydrogen-based fuels, potentially transporting energy from renewables over long distances—from regions with abundant energy resources to energy-hungry areas thousands of kilometres away" (What Is Green ... 2021). According to experts, the use of renewable energy sources and the increase in energy efficiency may reduce the CO_2 emission by 90% (Energy transition, https://www.irena.org/energytransition). It is also indicated by the projections of decreasing emissions of this gas by 2050 after the most desirable scenarios of implementing sustainable development, especially the zero-carbon economy option. They were presented in the World Energy Outlook 2021 (IEA 2021).

The increase in the importance of renewable energy in the structure of energy production and consumption globally has an additional advantage. It will allow for supplying energy even to the least privileged and poor people who so far have been living without access to traditional energy. In the twenty-first century, 1 billion people still do not have access to electricity. It affects mainly rural areas. Of the 20 countries where this problem is most acute, as many as 19 are in sub-Saharan Africa. In this part of the continent, the electricity supply has failed to improve and even deteriorated in recent years. Scientists indicate that green energy is a chance for such areas (Rawat and Sauni 2015; Kumar and Majid 2020; Gharaibeh et al. 2021).

It must be remembered that energy transition requires a full transformation of several systems—energy, economic and social ones—in the entire world economy. It cannot be replaced by the redevelopment of energy systems in the countries currently regarded as leaders of this transition. In 2021 the World Economic Forum's experts prepared the 10th edition of the report entitled Fostering Effective Energy Transition. 2021 Edition (World Economic Forum 2021). It evaluates the efficiency of energy systems in 115 countries. In order to build a synthetic indicator called Energy Transition Index, partial indexes were used. They refer to such areas as security and access to energy, sustainable development as well as economic growth and development.

In 2021 the first ten places in this ranking were held by: Sweden, Norway, Denmark, Switzerland, Austria, Finland, Great Britain, New Zealand, France and Iceland. Among them, European countries dominated. Within 10 years, only 13 out of 115 countries noted the improved results of this ranking. In the case of the world's remaining countries, the improved results of energy system efficiency within the last 10 years were very insignificant, or the index was dropped. The report results show that energy transition was unsystematically implemented in this period. Positive changes did not take place in all dimensions of transition. Such dimensions include capital and investments, infrastructure and innovative business environment, energy system structure, regulations and political obligations, institutions and management, as well as human capital and the percentage of consumers. In the countries with varying degrees of development, different results in striving for changes in systems were achieved. It indicates that different countries choose different energy transition goals and pathways to reach their targets and build a low-carbon economy. However, given global warming, strategies to achieve unreasonably modest targets for increasing the share of green energy in total energy consumption will not be able to halt climate change quickly enough to threaten human and planetary survival.

3 Energy Transition Under the Pressure of War

Until recently, it seemed that the transition to green energy would proceed by 2050 according to the stages and at the pace planned in the Paris Agreement (UN 2015), although experts have pointed out that it is proceeding too slowly. After Russia's aggression against Ukraine on 24th February 2002, a completely new era began in the history of Europe because the period of 75 years of peace ended. Furthermore, geopolitical, economic, social and technological conditions in the European continent and the world changed. The armed conflict and sanctions imposed on Russia (embargo on exports and imports, financial, transport, and personal sanctions) changed the dynamics of energy transition in Europe and the use of hydrocarbons from Russia.

Cutting off natural gas supplies to Poland by Gazprom since 25.04.2022 has become another factor which significantly changed the conditions of energy transition in Europe. It was caused by Poland's refusal to pay for gas in roubles. According to agreements with the European Union, Poland was to pay for gas in euros or dollars. It resulted from the necessity of remaining financial sanctions against Russia in force. Bulgaria was also affected by the gas supply cutting-off. Both countries served a transit function in transmitting Russian gas to Western Europe. Therefore, it can be expected that Russia's action will impact the gas market in entire Europe and lead to gas shortages.

European countries depend on Russian gas to varying degrees within the so-called energy mix (in energy production and consumption by energy carriers). In the case of Hungary, it holds as many as 25%, Germany—4%, Bulgaria—12% and Poland—11%. However, it only amounts to 1% in this structure for Belgium (Pokharel and Thompson 2022). It indicates that Russia uses natural gas as a political weapon to influence Ukraine's allies (Faisal 2022). In this situation, the pressure for Europe's energy transition and shift to green energy is even stronger.

The stages of providing climate neutrality and the manners of obtaining it, which were written down in the strategy of the European Union entitled the European Green Deal (EC 2019), will have to be changed in these new conditions. However, a new way to achieve this goal in 2050 has not been established yet. The Union and its members are verifying the assumptions concerning the reduction of greenhouse gas emissions by 55% by 2030 as well as the entire climate neutrality by 2050 (Łącka and Wojdyła 2022). The current geopolitical situation indicates that the European Union should use its renewable energy sources to a greater extent within the energy transition framework. It is an alternative strategy to become independent of fossil fuels imported from Russia. It allows for avoiding the situation in which other countries will deliver supplies of those raw materials harmful to the environment (e.g. Saudi Arabia, Qatar). It will increase the energy security of individual member countries and reduce inflationary pressures, which are to some extent the result of constant and increasing increases in gas, oil and electricity prices.

The IRENA experts, in their recent elaboration of World Energy Transitions Outlook 2022, stated that "Compounding crises underscore the pressing need to accelerate the global energy transition. Events of recent years have accentuated the cost to the global economy of a centralised energy system highly dependent on fossil fuels. Oil and gas prices are soaring to new highs, with the crisis in Ukraine bringing new levels of concern and uncertainty. The COVID-19 pandemic continues to hamper recovery efforts, while citizens worldwide worry about the affordability of their energy bills. At the same time, the impacts of human-caused climate change are increasingly evident around the globe" (IRENA 2022).

4 The Importance of Renewable Energy in the European Union

Green energy from renewable sources has been increasing its share in the structure of energy production and consumption in EU countries over the last several years. Between 2004 and 2018, the share of renewable energy in the Community almost doubled. The rate of contribution of renewable energy sources to gross final energy consumption (total energy demand in a region or country) in the EU in 2004 was 9.6%, and in 2018—18.9%. As far as recent years are concerned, they have shown an even faster pace of increasing this index. According to the data provided by Eurostat, in 2020, as many as 37%, i.e. almost two-fifth of electricity consumed in the European Union, was produced from renewable sources (Renewable Energy on the Rise… 2022). It is fostered by the climate, energy, and even scientific-research policy of the EU and decreases the costs of solar and wind energy.

Obviously, not all states belonging to the European Union show the same involvement in the development of green energy. Analysing the data concerning energy use from renewable sources, presented in Fig. 1, one can notice that Austria was a leader among member states. In 2020 it covered 78.2% of the total demand for electricity using renewable sources. Sweden comes second with the share of using green energy amounted to ca. 75%. The third place is held by Denmark (65%), and the fourth one belongs to Portugal (58%). In the group of "new" countries that joined the European Union, only Croatia and Latvia could show significant energy use from renewable sources. In 2020 this index for both countries amounted to 53%. More than half the countries (15 states) belonging to the EU obtained an index below the EU average. The lowest share of energy from renewable sources in the total energy consumption was noted in such countries as Malta (9.5%), Hungary (12%), Cyprus (12%), Luxembourg (14%) and the Czech Republic (15%).

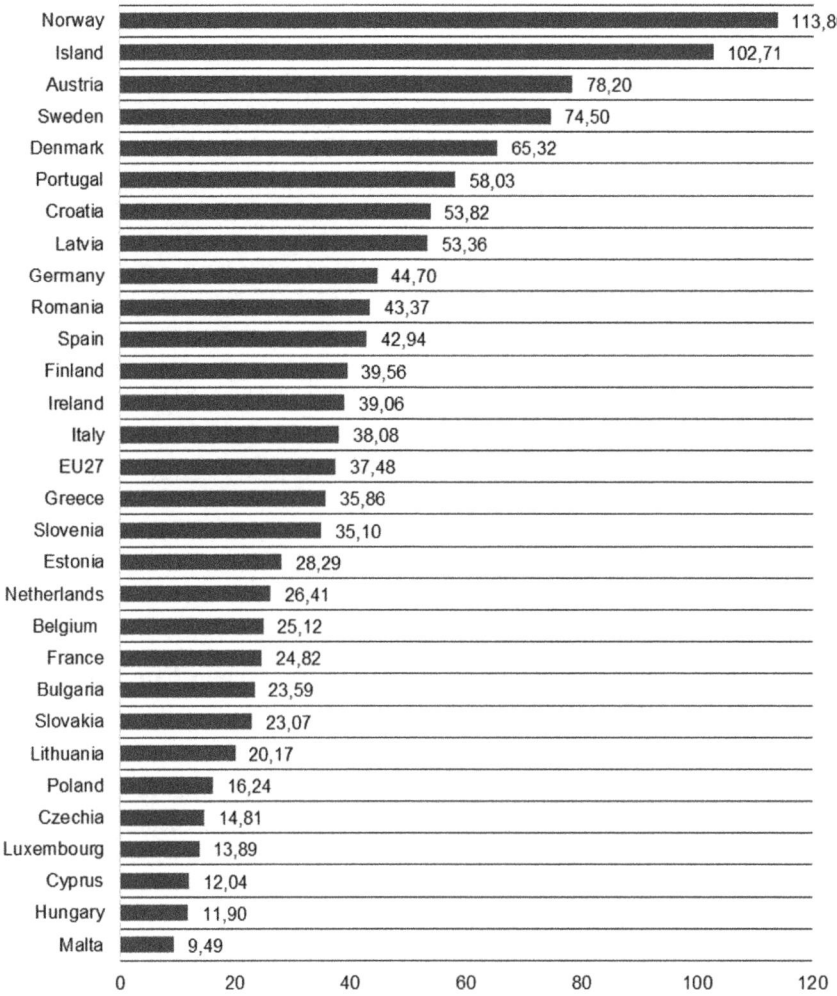

Fig. 1 The use of green energy in the EU countries in 2020 (%). *Source* Eurostat (2022); Renewable Energy on the Rise: 37% of EU's Electricity (2022)

At the same time, according to the data in Fig. 1, in 2020, Norway and Iceland (belonging to EFTA) reached a significantly higher share of the use of green energy (more than 100%) than other European countries. They produced more electricity from renewable sources than their total electricity consumption.

Eurostat data also show that wind (36%) and hydro (33%) sources accounted for most green energy in the European Union. In total, this accounted for more than two-thirds of total electricity from renewable sources. In contrast, solar energy (14%), solid biofuels (8%) and other renewable sources (8%) accounted for a smaller share.

A lower than EU average share of renewables in a country's total energy consumption in some cases may indicate a heavy reliance on fossil fuels (coal, gas and oil), but in others, the picture is not clear. Some countries may have other sources of energy which do not encumber the environment with a high emission of greenhouse gases and other substances harmful to nature and humanity.

Low carbon energy sources (emitting less CO_2 and dust and chemical substances responsible for smog) include the force of waters, waves and tides, geothermal energy, solar radiation, wind force, biomass and nuclear energy. According to the analysis of those various low carbon sources in energy production in the EU countries in 2020, for example, France, Slovakia and Belgium have a smaller share of renewable energy sources in the structure of total energy consumption than the EU average. At the same time, these countries have a substantial share of nuclear energy in the structure of energy production. France is the largest nuclear energy producer as it manufactures 52% of the total nuclear energy in the European Union. At the same time, this country is the most dependent on this type of energy. In 2020 nuclear energy constituted 70.6% of the entire electricity produced in France. In this period in Slovakia, over half of energy was produced in nuclear power plants (53.1%), and in Belgium, this index amounted to 39.1% (Statista 2022).

5 Hydrogen Technologies as a Support for Green Energy Development

During the research on electricity production technologies neutral for the environment (with the use of renewable energy sources), the concept of using hydrogen to produce electricity and heat emerged. This gas occurs widely in nature and can be generated artificially. Then ideas of using hydrogen as buffer fuel for energy storage appeared.

It turned out that hydrogen is a gas of a multifunction and multifaceted nature. On the one hand, it can be used as an energy carrier or a raw material. On the other, thanks to it, it is possible to store the energy generated seasonally from renewable sources. It is particularly important because renewable energy supplies are unstable, so there may be periodic (seasonal) shortages or large surpluses that national energy systems cannot absorb.

As Kłaczyński and Beroud (2021) indicate, almost 30% of produced green energy is lost in this case. Due to this fact, the current use of renewable energy is the most efficient in local areas, in the production places, where its use regularly is the easiest. Unfortunately, this does not solve the problems of providing the energy in large agglomerations for households, enterprises (particularly those that run a continuous production), and transport. The use of hydrogen as energy storage from renewable sources can ensure the stability of its supply while reducing the cost of production. Hydrogen has a great potential in terms of decarbonisation of the economy in the case of energy-intensive and high carbon-emitting industrial sectors (e.g. steel and fertiliser sector). This gas may also be used as the green fuel in heavy and collective transport. In those cases, existing battery technologies are impractical (IRENA 2019).

According to the information of the European Commission, nowadays in Europe, less than 2% of the energy use comes from hydrogen, and it is mainly used for the production of chemicals—plastic materials and fertilisers. Furthermore, it is applied in refineries, electronics, glassmaking, metal, and even the food industry. "In the petrochemical industry, it is used for hydrotreatment, hydrocracking, and reforming. The chemical industry uses hydrogen, inter alia, to produce ammonia, produce margarine by hydrogenating the unsaturated fats in vegetable oils, reduce iron ores, and to produce hydrogen chloride by combining chlorine and hydrogen (Häussinger et al. 2011). In the food industry, it is used as a food additive (E949), as it protects packaged goods against oxidation. It is also used in burners for cutting, soldering, and welding metals as well as fuel in hydrogen cells, internal combustion engines, and jet engines. It was originally used to fill balloons and airships. It can also be used as a cooling medium, for example, in turbo generators in power plants" (Gawlik and Mokrzycki 2021: 2).

Unfortunately, as much as 96% of this hydrogen is so-called grey hydrogen. It is produced from natural gas, which produces significant CO_2 emissions. The second type of hydrogen is blue hydrogen. It is made from natural gas, but CO_2 emissions from this process are reduced by using carbon capture methods. Next, it is stored underground or reused. This type of hydrogen is considered low-carbon and can be used to replace grey hydrogen in the interim. The most desirable, however, is green, renewable hydrogen. It is created by passing renewable energy (wind or solar power) through an electrolyser. Water is the only by-product of this process. Producing hydrogen in this way is almost emission-free (Gawlik and Mokrzycki 2021). Thus, it meets the European Green Deal's climate neutrality criteria and requirements. Due to this fact, in 2020, the European Union adopted a hydrogen strategy specifying necessary actions to take in order to develop the hydrogen economy, which will provide climate neutrality for the European Union countries in 2050 (EC 2020).

The strategy consists of three stages which are spread over time and lead to the establishment and increase in the scale of value chains based on the green hydrogen production. In the short term to 2024, these activities were expected to bring into operation electrolysers with a total capacity of 6 GW within the EU. They were to produce up to 1 million tonnes of green hydrogen used for the existing applications, i.e. the production of fertilisers in the chemical sector.

In the medium term to 2030, the plan is to increase the rate of construction of electrolysers and use hydrogen in new applications—as an energy carrier in energy-intensive industries (e.g. steel, fertilisers) and transport. It is assumed that the total capacity of electrolysers will increase to 40 GW. It was supposed to ensure effective annual production of up to 10 million tons of hydrogen.

In the long term, in the years 2030–2050, the green hydrogen production technology was to reach full maturity and spread over all areas of applications in which it is technically feasible and provides economic viability. It means that hydrogen technology was expected to demonstrate a cost advantage compared to alternative environmental technologies. The EU hydrogen strategy assumed that these technologies would be widely used in the entire EU, particularly in those industrial sectors that use processes difficult to electrify (Wolf and Zander 2021).

After Russia's attack on Ukraine and under the influence of the war (probably of a long-lasting nature), the European Union must quickly re-evaluate the assumptions described above about the pace of change, the technologies used and, above all, the role of gas as a transition fuel in the energy transition process. The planned process of building a hydrogen economy in Europe did not consider energy security and the necessity of resigning from natural gas in the conflict with Russia, the leading supplier of this raw material to the EU. Experts indicate that the share of green energy in providing energy security must be increased. Obviously, the use of fossil fuels from countries other than Russia is also necessary until a climate-neutral economy is established as soon as possible.

These problems constitute a strong incentive to intensify works on developing a hydrogen economy because the demand for green hydrogen will increase much faster than it was evaluated at the beginning of 2022. The original assumptions of hydrogen demand growth in the coming decades, both within the EU and in the global hydrogen market, are becoming outdated. Expert assessments of the pace of hydrogen technology development are also outdated. At the same time, the instability of the political situation and the prolonged armed conflict make it difficult to evaluate the current and medium- and long-term perspective of the hydrogen economy due to significant uncertainty and the risk of decision making (Łącka and Wojdyła 2022).

6 Advantages and Disadvantages of Green Energy

Energy production from renewable sources has many advantages and benefits the economy and its inhabitants. It is a clean energy source, and its use does not lead to the exploitation of limited resources of fossil fuels (gas, crude oil) or wood. Green energy production does not emit any pollutants in the form of CO_2 or other chemical substances in the form of dust. Thanks to the increase in using green energy, the quality of air and environment and the health of inhabitants in a given country are improved. The national emission of sulphur dioxide (SO_2), nitrogen oxides (NO_x), and inhaled fine particles with a diameter of 2.5 μm or smaller (PM2.5) is reduced. It reduces premature mortality from respiratory diseases and cancer and the incidence of allergies and reduces health care costs.

Furthermore, the advantages of green energy development include the reduced emission of greenhouse gases and decreased damage related to climate change. The use of renewable energy sources also limits water use for electricity production (Mai et al. 2016).

The development of energy technology and infrastructure related to renewable sources of energy fosters the development of the economy—the number of enterprises undertaking the investment in this area increases, the same as the number of workplaces in the development of renewable energy, supply chain, exploitation of green energy devices, the developing and newly created branches of the economy which are related to technologies of renewable sources of energy as well as in sectors producing devices for the needs of renewable energy.

There is an increased demand for construction services related to the construction of new generation units and the modernisation of existing ones. The development of the renewable energy sector and its particular pillars (including hydrogen economy) leads to intensifying research and scientific works and innovations that require the involvement of the R + D potential of the private and public sectors. The increased demand for high-quality human capital will shift highly qualified employees from traditional to innovative sectors (Ignarska 2013; IRENA 2017). Mai et al. (2016) suggest that the increase in the number of jobs in the renewable energy sector will be balanced by reducing employment in other sectors of the economy. In stable geopolitical conditions and with a normally functioning global natural gas market, renewable energy development would lead to lower gas prices due to a decrease in demand for this raw material. In the current situation of the war between Russia and Ukraine, this benefit cannot be expected.

According to the research conducted by the experts of the International Renewable Energy Agency (IRENA 2017), the development of renewable energy affects the global GDP growth and prosperity. It resulted from the increase in income, consumption and investment, the growth of employment (both men and women), reduction of health expenditure (due to reduced air pollution) and the increase in education at the same time. The quality of life improved. The results of this research confirm the conclusions of experts from American research centres, although they examined the impact of renewable energy development on the American economy (Mai et al. 2016). Gharaibeh and the team reached the same conclusions (Gharaibeh et al. 2021).

In addition to these numerous advantages and benefits of developing green energy in the world and national economies, one must also realise the disadvantages of renewable energy sources. Some sources of this energy, e.g. tides or hot springs (geothermal energy), are constant and stable energy sources. However, others do not guarantee stable energy supplies due to dependence on weather conditions (e.g. wind, water in rivers or solar radiation). Instability of supplies, periodical lack or surpluses of renewable energy are currently indicated as significant problems that may be solved through green hydrogen for its storage.

The most often indicated disadvantages of renewable energy include the high cost of investment, which causes the non-profitability of undertakings without obtaining financial support. It can take the following forms: preferential credits, subsidies or certificate systems. In recent years the problem of high investment expenditure in highly developed countries has become smaller (although not in all renewable energy technologies). It affects the development of the sector and technologies, which causes the long-term costs of creating and implementing technology, and acquiring competent employees become lower (the so-called external economies of scale).

The growing popularity of renewable energy sources in various highly developed countries is causing economies of scale to take effect. It is a result of increasing prices of electricity and financial incentives applied by the state, as well as the rising environmental awareness of their inhabitants—unit costs of implementing technologies of green energy and its production decrease. However, the primary constraints to increasing wave, wind, geothermal, tidal, biomass, biogas and biofuel power generation and the use of green hydrogen are still very high and high investment costs. Another disadvantage attributed to renewable energy sources is the negative impact of some of them on the environment and people living in the immediate vicinity of these energy sources. This argument explains the reluctance of societies to use the energy from tides, wind, photovoltaic installations or water power plants to a greater extent (Zabłocki 2013).

7 Barriers to the Development of Green Energy

Despite the numerous benefits of renewable energy, pressure from international organisations (e.g., the UN, the EU), growing promotion of renewable energy development by governments, and increasing public awareness of green energy, outside of highly developed countries (especially in the EU), the share of renewable energy in the energy production structure is small. An analysis of the findings of various authors by the team of Moorthy et al. (2019) shows that several categories of barriers to renewable energy development and utilisation can be identified. They result in lower competitiveness of renewable energy as compared to traditional energy and limit the possibility of achieving large scale production of renewable energy (Nasirov et al. 2015). They include social, economic, technological and regulatory barriers. In particular categories, Moorthy et al. (2019) listed and discussed specific factors limiting the use of renewable energy (Table 1).

In their research, Moorthy and his team sought to determine the degree and direction of impact of specific barriers to renewable energy deployment. Their research results indicate that social, technological and regulatory barriers have a powerful and direct impact on the development of green energy. However, economic barriers, although their influence is strong, act indirectly. The analysis of this research team shows that the development and implementation of renewable energy sources into the economy are strongly influenced by political aspects, which are related to the discrepancy between the objectives of state policy and adopted strategic assumptions and the actual results achieved during their implementation. The representatives of the authorities often show a lack of understanding for the need to implement the adopted goals, signed international declarations in the field of building a sustainable economy. They undertake the implementation of the tasks and goals included in the strategies with delay in an incomplete way. It results from succumbing to current challenges and the need to adjust their responses to satisfy what they believe are the more important problems at a given moment. Furthermore, politicians often face pressure from various advocacy groups, trying to gain voters' support. Recently it has been seen, for example, in Central and Eastern European countries (e.g. Poland) in the case of inhibiting wind and photovoltaic energy and in the implementation of the hydrogen strategy (Kryczka 2021; Łącka and Wojdyła 2022).

Table 1 Classification of barriers to the renewable energy development

Categories of barriers	Factors and their effects
Social	• "Not in my backyard" syndrome (NIMBY)—deployment of renewable energy in the economy is usually acceptable, but not in the immediate surroundings (objection of single citizens, interest groups, political leaders, civil society and sometimes even environmental organisations) • Loss of other alternative income, e.g. from agriculture or tourism as a result of allocating significant areas of land for the renewable energy purposes, limitations of fishing as a result of the development of offshore wind farms • Lack of experienced professionals necessary for the development of technology and construction as well as the exploitation of green energy facilities • Lack of awareness in terms of environmental and financial advantages of deploying renewable energy (RE) due to lack of knowledge and information barriers • Insufficient knowledge of the green energy technology • Uncertainty concerning the financial viability of RE projects
Economical	• High initial capital cost for RE installation projects • Underdeveloped system of financial support from the private investors accepting a higher level of risk • Too few investors willing to develop renewable energy • Competition from fossil fuels, the use of which has been more profitable due to lower prices • Government grants and subsidies—in the case of undertakings related to green energy, they are lower than those allocated for conventional energy • A high degree of investment risk • High cost of initial capital and a long period of the return on investment • Omission of external effects in the economic account of profitability of investment in conventional energy, which contributes to their higher efficiency in comparison to RE projects
Technological	• Limited access to advanced technology and infrastructure necessary for the development of renewable energy (particularly in underdeveloped countries) • Difficulties in the integration of conventional infrastructure and green energy • Lack of exploitation and maintenance culture—causes lower energy efficiency of devices as a result of neglected maintenance and operational errors • Insufficient investments in R + D cause green energy technologies to be in the development stage and encumbered with high costs and risks • The complexity of renewable energy technologies forces the establishment of standards, procedures and guidelines which provide durability, reliability, stability and efficiency of new solutions and possibilities for their commercialisation on a large scale

(continued)

Table 1 (continued)

Categories of barriers	Factors and their effects
Regulatory	• Poor government policy reflected in lack of cohesion in the national policy corresponding to provisions in the strategies concerning energy transition, lack of cohesion of the policy of renewable energy development (recently also the hydrogen economy), unstable energy policy, lack of policy integrating green energy technologies with the global market, improper action of government agencies • Inadequate financial and fiscal incentives • Administrative hurdles—delays and limitations in granting exemptions, excessive bureaucracy and excessive length of administrative proceedings, as well as high costs of construction permits concerning RE projects • Lack of certifications (standards) allowing for verification of whether the equipment and parts for the RE installation imported from abroad are consistent with standards in a given country

Source Moorthy et al. (2019)

8 Summary

In the third decade of the twenty-first century, the world economy faces immense environmental, economic, social, technological, and geopolitical challenges. They give rise to the growing social awareness of environmental damage due to using the energy from hydrocarbons combustion. These challenges evoke a feeling of the constant threat to the future of humanity unless we change the existing paradigm of development and energy transition. Additionally, Russia's aggression against Ukraine and cutting off some countries from gas supplies from Russia indicate the threat to energy security. It affects not only Poland and Bulgaria but also other countries which, as a result of sanctions against the aggressor, have experienced shortages of this raw material and constantly rising prices of fossil fuels.

The solution to these problems lies in the intensive development of renewable energy worldwide. However, it requires the reduction or complete removal of numerous barriers to renewable energy. In highly developed countries, they have less impact on the development of green energy. Yet, in many countries of the world, they are a significant constraint on the development of renewable energy technologies and the implementation of its solutions to meet energy demand. These barriers mean that today 1 billion people worldwide do not have access to energy. It limits the possibilities of social and economic development, especially in underdeveloped countries and the improvement of the quality of life of their inhabitants.

References

Bąk I, Cheba K (2020) Zielona gospodarka jako narzędzie zrównoważonego rozwoju. CeDeWu, Warszawa

Carayannis EG, Barth TD, Campbell DF (2012) The Quintuple Helix innovation model: global warming as a challenge and driver for innovation. J Innov Entrep 1:1–12. https://doi.org/10.1186/2192-5372-1-2

Dacko M, Jakubik-Grzybowska M, Łącka I, Malkowski A, Oleńczuk-Paszel A, Płonka A, Śpiewak-Szyjka M (2020) Ochrona środowiska i przyrody - wybrane aspekty prawne i ekonomiczne dotyczące JST. C.H. Beck, Warszawa

EC (2019) The European Green Deal. https://ec.europa.eu/info/sites/default/files/european-green-deal-communication_en.pdf. Accessed 15 April 2022

EC (2020) A hydrogen strategy for a climate–neutral Europe, Communication from the Commissions to the European Parliament, the Council The European Economic and Social Committee and the Committee of the Regions. https://ec.europa.eu/energy/sites/ener/files/hydrogen_strategy.pdf. Accessed 11 April 2022

EC (2010) European 2020. A European strategy for smart, sustainable and inclusive growth. https://ec.europa.eu/eu2020/pdf/COMPLET%20EN%20BARROSO%20%20%20007%20-%20Europe%202020%20-%20EN%20version.pdf. Accessed 15 April 2022

Energy Transition (2022). https://www.irena.org/energytransition. Accessed 15 April 2022

Eurostat (2022) Data browser. Share of energy from renewable sources. https://ec.europa.eu/eurostat/databrowser/view/NRG_IND_REN. Accessed 21 April 2022

Faisal I (2022) Ukraine war: Russia halts gas exports to Poland and Bulgaria. https://www.bbc.com/news/business-61237519. Accessed 27 April 2022

Gawlik L, Mokrzycki E (2021) Analysis of the Polish hydrogen strategy in the context of the EU's strategic documents on hydrogen. Energies 14:6382. https://doi.org/10.3390/en14196382

Gharaibeh AA, Al-Shboul DA, Al-Rawabdeh AM, Jaradat RA (2021) Establishing regional power sustainability and feasibility using wind farm land-use optimization. Land 10:442. https://doi.org/10.3390/land1005044

Häussinger P, Lohmüller R, Watson AM (2011) Hydrogen 6. Uses. In Ullmann's Encyclopedia of industrial chemistry. Wiley, VCH, Weinheim

Henderson J, Sen A (2021) The energy transition: key challenges for incumbent and new players in the global energy system. OIES Paper: ET 01, Oxford Institute for Energy Studies. https://www.oxfordenergy.org/publications/the-energy-transition-key-challenges-for-incumbent-and-new-players-in-the-global-energy-system/ Accessed 4 May 2022

Hoegh-Guldberg O, Jacob D, Taylor M, Bindi M, Brown S, Camilloni I, Diedhiou A, Djalante R, Ebi KL, Engelbrecht F, Guiot J, Hijioka Y, Mehrotra S, Payne S, Seneviratne SI, Thomas A, Warren R, Zhou G (2018) Impacts of 1.5 °C global warming on natural and human systems. In: Global warming of 1.5 °C. An IPCC special report on the impacts of global warming of 1.5 °C above pre-industrial levels and related global greenhouse gas emission pathways, in the context of strengthening the global response to the threat of climate change, sustainable development, and efforts to eradicate poverty (in press). https://www.ipcc.ch/site/assets/uploads/sites/2/2019/06/SR15_Chapter3_Low_Res.pdf. Accessed 3 May 2022

ICC (2012) Green Economy Roadmap and Ten Conditions for a Transition toward a Green Economy. https://iccwbo.org/content/uploads/sites/3/2012/08/Green-Economy-Roadmap-a-guide-for-business_-policy-makers-and-society.pdf. Accessed 21 April 2022

IEA (2021) World Energy Outlook. https://www.iea.org/reports/world-energy-outlook-2021. Accessed 7 May 2022

Ignarska M (2013) Odnawialne źródła energii w Polsce. Poliarchia 1(1):57–72. https://doi.org/10.12797/Poliarchia.01.2013.01.06

IRENA (2017) Renewable energy benefits: understanding the socio-economics. https://www.irena.org/-/media/Files/IRENA/Agency/Publication/2017/Nov/IRENA_Understanding_Socio_Economics_2017.pdf. Accessed 1 May 2022

IRENA (2019) Hydrogen: a renewable energy perspective. https://www.irena.org/-/media/Files/IRENA/Agency/Publication/2019/Sep/IRENA_Hydrogen_2019.pdf. Accessed 12 April 2022

IRENA (2022) World Energy Transitions Outlook 2022: 1.5 °C pathway. https://www.irena.org/publications/2022/Mar/World-Energy-Transitions-Outlook-2022 Accessed 26 April 2022

Jing E (2016) Development of renewable energy in Australia and China: a comparison of policies and status. Renew Energy 85(C):1044–1051. https://doi.org/10.1016/j.renene.2015.07.060

Johansson TB, McCormick T, Neij L, Turkenburg WC (2006) The Potentials of Renewable Energy. In: Assmann D (ed) Renewable energy. A global review of technologies, policies and markets, eBook Published. https://doi.org/10.4324/9781849772341

Kłaczyński R, Beroud A (2021) Wykorzystanie wodoru jako alternatywnego źródła pozyskiwania energii w strategii energetycznej Federacji Rosyjskiej. Colloquium 2(42):57–70. https://doi.org/10.34813/14coll2021

Kryczka D (red) (2021) Rozwój gospodarki wodorowej w UE i państwach członkowskich – środowisko regulacyjne i finansowe. ESPERIS, UN Global Compact, Warszawa

Kumar JCR, Majid MA (2020) Renewable energy for sustainable development in India: current status, future prospects, challenges, employment, and investment opportunities. Energ Sustain Soc 10:2. https://doi.org/10.1186/s13705-019-0232-1

Łącka I, Wojdyła P (2022) Transformacja energetyczna w cieniu wojny. Finansowanie ze środków unijnych szansą na rozwój technologii wodorowych w Polsce. Chem Przemysłowa 2:72–78

Łuniewski A, Łuniewski S (2020) Środowiskowe i ekonomiczne aspekty zielonej gospodarki w regionach przygranicznych Polski i Białorusi. Oficyna Wydawnicza Politechniki Białostockiej, Białystok

Mai T, Wiser R, Barbose G, Bird L, Heeter J, Keyser D, Krishnan V, Macknick J, Millstein M (2016) A prospective analysis of the costs, benefits, and impacts of U.S. Renewable Portfolio Standards. NREL/TP-6A20-67455/LBNL1006962. National Renewable Energy Laboratory and Lawrence Berkeley National Laboratory, Golden, CO and Berkeley. http://www.nrel.gov/docs/fy17osti/67455.pdf. Accessed 4 May 2022

Moorthy K, Patwa N, Gupta SY (2019) Breaking barriers in deployment of renewable energy. Heliyon 5(1):e01166. https://doi.org/10.1016/j.heliyon.2019.e01166

Mrozowska S, Wendt JA, Tomaszewski K (2021) The challenges of Poland's energy transition. Energies 14:8165. https://doi.org/10.3390/en14238165

Nasirov S, Silva C, Agostini CA (2015) Investor' perspectives on barriers to the development of renewable energy sources in Chile. Energies 8(5):3794–3814. https://doi.org/10.3390/en8053794

Pokharel S, Thompson M (2022) Russia shuts off gas supplies to Poland and Bulgaria. https://edition.cnn.com/2022/04/26/energy/poland-russia-gas/index.html. Accessed 27 April 2022

Rawat D, Sauni P (2015) Importance and prospects of renewable energy: emerging issue in India. Int J Art Hum Sci 2(4):11–18

Renewable Energy on the Rise: 37% of EU's Electricity (2022). https://ec.europa.eu/eurostat/web/products-eurostat-news/-/ddn-20220126-1. Accessed 2 May 2022

Statista (2022) Energy&Environment. Share of nuclear power in total domestic electricity generation in 2020, by selected country. https://www.statista.com/statistics/270367/share-of-nuclear-power-in-the-power-supply-of-selected-countries/. Accessed 4 May 2022

Tănasie AV, Năstase LL, Vochița LL, Manda AM, Boțoteanu GI, Sitnikov CS (2022) Green economy—green jobs in the context of sustainable development. Sustainability 14:4796. https://doi.org/10.3390/su14084796

Turkenburg WC (2015) Renewable energy technologies. World energy assessment: energy and challenges of sustainability. UNDP, New York, pp 219–272

UN (2015) Paris Agreement. https://unfccc.int/sites/default/files/english_paris_agreement.pdf. Accessed 22 April 2022

UN (2021) Theme report on energy transition. Towards the achievement of SDG 7 and net-zero emissions. https://www.un.org/en/hlde-2021/page/theme-reports. Accessed 22 April 2022

UNEP (2011) Towards a green economy: pathways to sustainable development and poverty eradication—a synthesis for policy makers. https://sustainabledevelopment.un.org/index.php?page=view&type=400&nr=126&menu=35. Accessed 7 May 2022

What Is Green Hydrogen and Why Do We Need It? An expert explains. https://www.weforum.org/agenda/2021/12/what-is-green-hydrogen-expert-explains-benefits. Accessed 27 April 2022

Wolf A, Zander N (2021) Green hydrogen in Europe: do strategies meet expectations? Intereconomics 6:316–323. https://doi.org/10.1007/s10272-021-1008-3

World Economic Forum (2021) Fostering effective energy transition, 2021 edn. Cologny/Geneva. https://www3.weforum.org/docs/WEF_Fostering_Effective_Energy_Transition_2021.pdf. Accessed 12.04.2022

Zabłocki M (2013) Determinanty wykorzystania odnawialnych źródeł energii w Polsce. Technika Poszukiwań Geologicznych. Geotermia. Zrównoważony Rozwój 2:29–44

Green Energy and Its Impact on Environmental Protection

Beata Szczecińska

Abstract In contemporary political debate and scientific considerations, the dominant view is that green transformation is a remedy for all the problems of the modern world. The standard of living in the world is increasing, but people's awareness of environmental protection is also increasing. In practice, however, it turns out that the introduction of green energy solutions to the market also poses various problems. The purpose of this discussion is to attempt to assess the environmental impact of green energy. The chapter presents issues related to the impact of green energy on environmental protection, both in terms of benefits and problems resulting from the widespread use of green technologies. The literature research suggests that every way of obtaining energy is not entirely environmentally friendly. However, the researchers agree that the benefits of green energy are much greater than the negative consequences, and the latter occurs mainly in the production, installation and disposal of energy devices. Low emissions of harmful gases to the atmosphere and no release of toxic pollutants into water reservoirs place renewable energy sources on the right side of environmental protection.

1 Introduction

The development of humanity is inextricably linked to the development of the economy. However, it has been known that human activities have a significant negative impact on the natural environment, threatening the existence of future generations. Economic growth is often the most important goal for many political leaders and entrepreneurs who do not fully believe in development opportunities without negative effects on the environment. However, increasingly widely disseminated media information reaches the public, contributing to increased awareness of environmental protection and sustainable development opportunities. Implementing sustainable development in individual regions or countries depends on many economic and

---

B. Szczecińska (✉)
Faculty of Economics, West Pomeranian University of Technology in Szczecin, Szczecin, Poland
e-mail: beata.szczecinska@zut.edu.pl

social factors. Furthermore, it is common knowledge that to make a change, one has to change the way of thinking in the first place.

In order to achieve the objective of sustainable development and promote the harmonious development of man and nature, the concept of green environmental protection should be applied. This concept cannot be implemented in the short term and should be considered long-term. The main content of the concept of environmental protection is the rational and sustainable use of certain natural resources and the integration of the idea of environmental sustainability and cleaner production. The development of the concept of environmental protection will allow reducing environmental pollution and the harmonious development of man and nature (Ding et al. 2018).

In contemporary political debate and scientific deliberations, the dominant view is that the green transition is the cure for all the problems of the modern world. In this chapter, issues related to the environmental impact of green energy will be presented from the perspective of both the benefits and problems of widespread adoption of green technologies.

The green economy, including green energy, assumes a positive environmental impact. Although many of its solutions really contribute to it, do they always? Will the positive effects on the environment observed at the moment be long-lasting? Will they contribute to improving conditions on Earth and stopping its degradation? The paper attempts to answer these questions, examine past research results, and assess the real situation. That is the main purpose of this discussion.

In the literature on the subject, the vast majority of publications and research results concern the adverse environmental effects of the use of non-renewable energy. However, fewer studies explicitly analyze the positive and negative effects of green energy on the environment—the considerations in this chapter attempt to bridge the gap in this area.

2 Energy Demand and the Environment

Economic growth is inextricably linked to an increase in energy demand. Energy consumption from both traditional and renewable sources has increased in the last few decades. However, the use of conventional energy sources has serious negative effects. Environmental degradation, pollution and global warming are spending their dream from the minds of scientists, researchers, politicians and economists. Although the environment offers various sources that can be used in industry and daily life, still many entrepreneurs focus on the economic effects, thus destroying the environment. Studies on the impact of economic growth and increasing energy demand on the environment were conducted, among other things, by Shahbaz et al. (2013), Fan et al. (2019a, b), Khan et al. (2019), Khan et al. (2022). The conclusions of their research were similar and clearly indicated the degradation of the environment caused by urbanization and industrial development.

The fact that the overconsumption of energy from traditional sources, such as fossil fuels, has led to global severe air pollution challenges for both developed and developing economies has been reported by Shen et al. (2020). They found that green renewable energy innovations reduced air pollution both locally (in the study area) and in neighbouring provinces. Therefore, the authors point to the need for environmental regulation.

The negative effects of unsustainable consumption of natural resources and pollutant emissions are felt practically in every country. Therefore, many researchers point to the need for systemic solutions to reduce the negative effects of increased energy consumption on the environment. Khan et al. (2022), based on the conclusions of the studies, proposed several helpful policy implications. First, politicians and government agencies must take decisive action to address the problem of air pollution from industry and road transport. It requires the construction of an energy-efficient and environmentally friendly transport infrastructure that will contribute to the reduction of carbon dioxide emissions. Secondly, it is important to ensure that the solutions implemented are long term to avoid further negative environmental impacts. Thirdly, enabling and facilitating the use of green technologies in manufacturing processes will affect the quality of the environment. Finally, state interference is crucial in developing specialized methods to counter and prevent environmental degradation by educating the public.

In view of the serious environmental situation, Peng et al. (2020) also wrote on the need for systemic solutions, giving examples of various measures implemented by individual countries to reduce environmental pollution. Shen et al. (2020) believe that in formulating environmental policies, emphasis should be placed on coordination and communication between regions.

Meeting the maintenance of energy efficiency on the one hand and the mitigation, even elimination of environmental problems on the other, is clearly the role of politicians. Today, environmental policies and taxes are used to promote green technologies, energy efficiency and a cleaner and healthier environment (Shahzad 2020). Environmental taxes and various deductions and subsidies impact the use of innovative technologies that reduce environmental pollution. They are important instruments in the hands of the Member States, which can have a tangible impact on protecting the environment.

Achieving the goals of a sustainable economy requires the use of green energy. Meeting the goals of a sustainable economy requires the use of green energy. Green energy projects (including wind, solar, biomass, and hydroelectric projects) are major components of biofuel projects and basic needs of the global world that directly affect economic growth and gross domestic product development. Over the last few decades, economic growth and the growing population have contributed to increasing fossil fuel consumption and carbon dioxide emissions (Dar et al. 2022). Wang and Prinn (2011) estimated that installing the optimal number of wind turbines in offshore waters less than 600 m deep worldwide could potentially meet up to 25% of the projected future global energy demand.

Everything affects the environment in some way; every energy source does the same. Dar et al. (2022) consider that understanding and managing global climate change depends on understanding how environmental pollution is linked to economic activity. Most studies show that petroleum derivatives such as coal, oil and flammable gas are much more destructive to air and water pollution than sustainable energy. Unlike non-renewable energy sources such as petroleum products, sustainable energy sources such as wind, sun, geothermal energy, biomass and hydro have fewer ecological consequences. The nature and strength of eco-effects vary depending on the technology used, location and other factors. As renewable energy sources account for a growing portion of the world's electricity supply, environmental consequences can be effectively prevented or reduced by knowing and understanding the existing and future environmental problems of each renewable energy source.

Based on a literature review, the following sections describe the advantages and disadvantages of the most popular green energy sources in relation to the environment.

3 Beneficial Effects of Green Energy on the Environment

The use of conventional technologies to generate energy has a destructive effect on the environment by enlarging the ozone hole, the greenhouse effect, acid rain, the accumulation of radioactive substrates and the occurrence of smog. All this also affects the health and life of people and other living organisms. The green economy, including the use of green energy, should help address global challenges such as climate change, biodiversity loss and desertification. It should also contribute to national and regional efforts to tackle local air, water and soil pollution. The transition to a green economy will also bring economic benefits (of course, any such change also entails risks and costs). An obvious potential advantage of the green economy is opening new export markets for biofuels and renewable energy technologies such as solar panels and wind turbines. In addition to opening up new markets, the transition to a green economy can help maintain market share (Cosbey 2011).

Investment in renewable energy technologies plays a crucial role in the fight against climate change and contributes to achieving carbon neutrality objectives (Yang et al. 2022). The clean or renewable energy sector is a tool for growth and innovation in all industries. Investing in clean energy not only leads to improved energy security and improved environmental and public health, but studies confirm additional benefits in the short term, such as employment growth opportunities (Direct Benefits of Green Economy 2020). Job creation is not only related to direct employment in companies dealing strictly with renewable energy but also in related sectors (manufacturing and services). Thus, it can improve the overall quality of life of people.

Analyses, both single and multinational, of energy consumption concerning environmental quality, have been conducted by many researchers. The results of many of these studies were collected and described by, for example, Shahzad (2020). They indicate a causal relationship between energy consumption from non-renewable sources (from fossil fuels) and high pollutant emissions (e.g. Wang et al. 2011; Ozcan 2013; Muhammad 2019).

Specific findings on the beneficial effects of green energy use on environmental pollution were obtained by Kumar and Madlener (2016). They attempted to investigate the role of renewable energy in reducing carbon dioxide emissions in India using the Long-range Energy Alternatives Planning (LEAP) method. Based on empirical analysis, the authors suggest that with the large-scale deployment of renewable energy technologies, which estimates that 23% of total electricity is produced from renewable sources, carbon dioxide emissions are expected to be reduced by 74% by 2050.

An attempt to assess the short- and long-term impact of renewable energy on carbon dioxide emissions into the atmosphere was also carried out by Zoundi (2017). The study covered 33 years (1980–2012) and 25 African countries, the selection of which was driven by data availability and significant activities in the introduction of renewable energy sources (hydro, geothermal, solar, wind and biomass)—the collected data allowed to analyze panel cointegration to determine long-term effects. The results show that renewable energy can meet household energy demand and improve air quality, with a 1% increase in renewable energy consumption, reducing CO_2 emissions by 0.13%. In contrast, a 1% increase in population leads to 0.026% additional carbon emissions (at a significance level of 10%). As a result of the rapid increase in transportation and demand for goods and services, this creates more pressure on resources and the environment. It is therefore recommended to switch to green energy whenever possible.

According to many authors, renewable energy sources have minimal adverse effects on the environment, are economical, have a fixed unit cost of energy obtained, and operate on a separate grid, so there is no need to transport the energy they produce (Kieć 2007). Unlike fossil fuels and nuclear power, Leung and Yang (2012) believe that wind power is environmentally friendly because wind turbines do not pollute our atmosphere with greenhouse gases, nor do they cause any problems with radioactive waste for future generations.

According to Hernandez et al. (2014), renewable energy is a promising alternative to fossil fuel-based energy, but its development may require a complex set of environmental trade-offs. The authors reviewed the direct and indirect environmental impacts—both beneficial and adverse—of industrial-scale solar energy development, including impacts on biodiversity, land use and land cover change, soils, water resources, and human health. In the summary of their research, they stated that solar energy has several positive aspects: reducing greenhouse gases, stabilizing degraded land, increasing energy independence and employment opportunities, accelerating the electrification of rural areas and improving the quality of life. As a result, solar power plants will have negligible direct effects that negatively affect biodiversity. According to Tsoutsos et al. (2005), photovoltaics generally have a mild impact on the environment, not generating noise or chemical pollution during operation. It is one of the most cost-effective renewable energy technologies used in urban environments.

Energy from green sources is becoming more widespread, although biomass has already been the basis of the global fuel economy by the mid-eighteenth century. Then fossil fuels were first placed, but recently there has been a growing interest in this energy source again due to its lower carbon dioxide emission into the atmosphere (Abbasi and Abbasi 2010). The fuel obtained in this way burns cleaner than fossil fuels. In addition, biomass can be used not only as a source of energy but also as chemicals added to medical and food products (Hamidpour et al. 2021).

Another source of green energy, already known for many years, is hydroelectric power plants, in which turbine-driven generators convert kinetic energy (from fast-flowing or falling water) into mechanical energy. They produce very few greenhouse gases, do not contribute to the increase in global temperature and are the least expensive way to store large amounts of electricity (Faizal et al. 2017).

Public information shows that energy from renewable sources is an environmentally friendly way of energy production. Companies producing and assembling green energy carriers, e.g. solar panels or electric windmills, praise them, giving mainly the benefits of their use. A look at any green energy company's website reveals the benefits of using such technologies. The positive impact on the environment and the improvement of the quality of life of people are emphasized.

The review of the scientific literature also outlines an optimistic vision of increasing the share of green energy carriers and their impact on mitigating global warming and improving the ecosystem. Renewable energy technologies are based on natural energy resources such as solar radiation, winds, and waves, which are constantly replenished and therefore will not run out. Available literature (Masson et al. 2014; Nugent and Sovacool 2014; Elliott 2022, among others) indicates that the benefits of green energy far outweigh the negative consequences, with the latter occurring mainly in the production and disposal of energy equipment. Low emissions of harmful gases to the atmosphere and no release of toxic pollutants into water reservoirs place renewable energy sources on the right side of environmental protection.

4 Disadvantages of Switching to Ecological Solutions in the Energy Sector

A literature review on various energy sources clarifies that no energy sector is fully environmentally friendly. Obtaining energy consists of many activities that have a greater or lesser negative impact on people and the natural environment. Each process, such as obtaining, processing, separating and transporting energy carriers to a different extent, is associated with a threat to living organisms (Ziębik and Szorgut 1997).

Wind farms are increasingly popular sources of green energy globally, especially in windy areas with specific terrain. Both onshore and offshore wind farms are sources of vibration noise and air vibration and can affect the environment differently. While most publications report little negative impact, the rise in popularity of their use somehow forces the need for an increasing number of in-depth studies in this area. The long-term effects are particularly interesting, although they are still not well known.

The negative impact of wind farms on the environment can be considered from the point of view of humans, birds and marine animals and climate change. Knopper and Ollson (2011), analyzing the results of various studies from the reviewed articles, concluded that there is no direct causal link between people living in the vicinity of modern wind turbines, the noise they emit, and the resulting physiological health effects. Similar conclusions were reached by Bolin et al. (2011). They believe that the prolonged low-frequency noise that accompanies a rotating turbine can indirectly cause headaches, sleep disturbances, hearing loss, irritation, anxiety, and even stress symptoms. It can also harm the vestibular system (Punch et al. 2010). Therefore, it is suggested to put turbines at a distance of 2 km from residential buildings or build obstacles (sound-absorbing curtains) on the way of propagation. The visual impact of turbine operation on people is also noted, although assessed very subjectively. Some find wind turbines impressive and enjoyable, while others have opposing views (Leung and Yang 2012). The danger of birds killing themselves over wind turbines was described by Hau (2006) in his book. However, he found that birds can quickly get used to a new situation and learn to avoid dangerous obstacles. The impact of wind farms on marine animals is not well understood. However, few studies indicate that some species may be sensitive to them, especially during their construction and hammering foundations (Mann and Teilmann 2013).

Local climate change, caused by the operation of wind turbines, is also not thoroughly studied, and it is even more difficult to determine whether this impact is negative or positive. Hau (2006) believes that measurable impacts on environmental climate are possible, if any, only with a massive system of large wind turbines. However, the author doubts the negative climate effects of energy so harvested. Research by Wang and Prinn (2011) has shown that massive deployment of offshore wind, covering a significant portion of total global energy consumption, will result in a slight (about 0.2°) cooling of the ocean surface in areas with wind farms. In contrast to the situation on land, earlier studies by the same authors showed a warming of the land surface (Wang and Prinn 2010). The authors point out that the results of similar studies may differ from each other due to the research method used. Mann and Teilmann (2013) also stress that it is difficult to transfer the results of the impact of wind farms from one location to another and encourage further research.

Industrial-scale solar power plants require large land areas, so their location must be carefully considered. They change the landscape to a large extent due to the need for site preparation, construction of access roads and necessary infrastructure. These construction activities cause dust emissions and generate noise. In addition, solar energy devices require significant site preparation (including the removal of vegetation), which changes the topography and thus the ability of the soil to conduct water (Abbasi and Abbasi 2000). These factors can have a negative impact on the environment. The operation of the finished energy equipment is relatively quiet, so it will not affect the health and life of the fauna. Sounds are only generated by pumps and fans, and the problem may occur as the number of fans increases due to the increase in power generation (Lovich and Ennen 2011). Although the direct impact on living organisms in the case of solar power plants is defined as negligible, it should be borne in mind that indirectly the environmentally toxic substances required for the operation of the equipment (e.g. dust agents, rust inhibitors, antifreeze) and herbicides can have unhealthy and potentially long-term consequences for local biodiversity (Abbasi and Abbasi 2000).

Trimble et al. (1984) wrote about the fact that biomass is a renewable fuel with low sulfur content, which, however, is not without its environmental impact. Among the problems they mentioned was competition for arable land necessary for food production, depletion of nutrients from the soil and deterioration of water quality. In addition, the biochemical processes of converting biomass products into fuel cause air pollution. These effects can be mitigated by planning and applying appropriate environmental control technology. Problems between biomass energy and the environment have also been discussed by Mao et al. (2018), Wu et al. (2018), Al-Shetwi (2022). They pointed out that bioenergy production can indeed have a negative impact on the environment in terms of water quantity and quality, greenhouse gas emissions, biodiversity and organic carbon in the soil, and soil erosion.

Water energy contributes to meeting the growing demand for clean energy but is also controversial. Interference with the environment in constructing this type of power plant is undeniable. Alteration of the river's hydrology by making the land slope for dam construction can significantly reduce aquatic biodiversity and migratory fish populations (Ligon et al. 1995; Magilligan and Nislow 2005; Yu et al. 2019). In addition, the construction of power plants may be accompanied by regional deforestation and displacement of people (Moran et al. 2018). According to Faizal et al. (2017) and Moran et al. (2018), possible negative environmental impacts can be avoided by building small hydropower plants.

Direct effects on living organisms can also occur outside the location of energy facilities. The extraction of large quantities of raw materials for the construction of renewable energy facilities (e.g. aggregate, cement, steel, glass), the transport and processing of these materials, the need for large amounts of water to cool specific installations and the generation of toxic waste, including coolants, antifreeze, rust inhibitors and heavy metals, may affect the environment (Abbasi and Abbasi 2000; Tsoutsos et al. 2005; Lovich and Ennen 2011).

However, closer encounters with green energy make it possible to list the occasional yet negative consequences of its use. The results of many studies (e.g. Chien et al. 2021) confirm that both in the short and long term, renewable sources do not adversely affect environmental degradation to the same extent as non-renewable energy sources (especially fossil fuels).

By developing environmental legislation for a low-carbon energy system, countries can encourage industry to use green renewable and sustainable energy technologies. By using this approach, countries can continue to implement industry-level policies to provide incentives and subsidies for adopting environmentally friendly technologies that can help stimulate sectoral innovation to combat climate change problems (Shahzad 2020). However, in most cases, the cost of such new technologies and regulations is not assessed.

5 Summary

Consumption of natural resources without restraint results in constant deterioration of natural conditions; therefore, all possible green economy solutions should be tested and implemented. Thanks to this, there is a chance for people to live in comfortable, non-life-threatening conditions.

The climate problem is not caused by economic growth but by the lack of adequate public policies to reduce greenhouse gas emissions. Clean air and water, healthy food and preserved nature benefit human health and also bring far more economic benefits than economic costs (Cohen 2020).

Many theoretical and empirical publications emphasize the negative impact of non-renewable energy use on the environment. The main conclusion of this work is that the use of green energy will stop the degradation of our planet. Unfortunately, while the development of renewable energy technologies is relatively simple (though still costly), problems with social and institutional implementation are often much more difficult to solve. The resistance of many social groups to change is still noticeable and problematic. Therefore, in the first place, each of us should learn to save energy in every possible place: at home, at work, and on the streets, which will reduce the energy demand. Public awareness of this issue is changing (especially in the younger generation), and many people are trying to make radical changes in their lives, but it is still not enough. Therefore, the necessity of introducing systemic solutions concerning pollution control, allowances and facilities connected with the use of renewable energy sources, which is emphasized in many studies, may make it possible to increase the share of these carriers in the structure of all the ways of energy production used. The experience of many countries shows the positive effects of using pollution control technologies in compliance with local laws. Of course, regulations for enforcing these laws vary, so the environmental benefits also vary.

Politicians still face the problem of reconciling economic development with environmental protection because many people, mainly from highly developed countries, cannot imagine giving up the facilities of this world and changing their lifestyles.

Energy production is a critical issue for every country, so many people (scientists, advisors, politicians, entrepreneurs) are concerned with this issue. Without energy, there is no life, but energy production also takes life, so it is not surprising that people are looking for ways to produce energy that would have the least invasive impact on the environment. This chapter refers to the results of many scientists' research, but many important studies are not referred to here. Thereby, this confirms the undeniable fact that the topic is very important, timely, and requires further research in the search for answers to the question of the future of our planet.

References

Abbasi SA, Abbasi N (2000) The likely adverse environmental impacts of renewable energy sources. Appl Energy 65(1):121–144. https://doi.org/10.1016/S0306-2619(99)00077-X

Abbasi T, Abbasi SA (2010) Biomass energy and the environmental impacts associated with its production and utilization. Renew Sustain Energy Rev 14(3):919–937. https://doi.org/10.1016/j.rser.2009.11.006

Al-Shetwi AQ (2022) Sustainable development of renewable energy integrated power sector: trends, environmental impacts, and recent challenges. Science Total Environ 822. https://doi.org/10.1016/j.scitotenv.2022.153645

Bolin K, Bluhm G, Eriksson G, Nilsson ME (2011) Infrasound and low frequency noise from wind turbines: exposure and health effects. Environ Res Lett 6(3). https://doi.org/10.1088/1748-9326/6/3/035103

Chien F, Sadiq M, Nawaz MA, Hussain MS, Tran TD, Le Thanh T (2021) A step toward reducing air pollution in top Asian economies: the role of green energy, eco-innovation, and environmental taxes. J Environ Manag 297. https://doi.org/10.1016/j.jenvman.2021.113420

Cohen S (2020) Economic growth and environmental sustainability. State of the Planet, Climate, Earth, and Society, Columbia Climate School. Available from: https://news.climate.columbia.edu/2020/01/27/economic-growth-environmental-sustainability/?msclkid=7f2da506b64d11ec98a7108cbd972e9d. Accessed 05 April 2022

Cosbey A (2011) Trade, sustainable development and a green economy: benefits, challenges and risks. In: Transition to a green economy: benefits, challenges and risks from a sustainable development perspective. United Nations Department of Economic and Social Affairs (UNDESA). Available from: https://wedocs.unep.org/bitstream/handle/20.500.11822/9310/-Transition%20to%20a%20green%20economy:%20benefits,%20challenges%20and%20risks%20from%20a%20sustainable%20development%20perspective-2012UN-DESA,%20UNCTAD%20Transition%20GE.pdf?sequence=3&amp%3bisAllowed=&msclkid=fd6f422db3ff11ecac8ab8f0c98a38e1. Accessed 02 April 2022

Dar AA, Hameed J, Huo C, Sarfraz M, Albasher G, Wang C, Nawaz A (2022) Recent optimization and panelizing measures for green energy projects; insights into CO_2 emission influencing to circular economy. Fuel 314. https://doi.org/10.1016/j.fuel.2021.123094

Ding Y, Zhang X, Liu Q et al (2018) Analysis of green environmental protection concept in environmental engineering design. Smart Constr Res 2(1). https://doi.org/10.18063/scr.v0.425. Available from: https://www.researchgate.net/publication/325188966_Analysis_of_green_environmental_protection_concept_in_environmental_engineering_design?msclkid=e22a89b2b58e11ec9744f5b9ddec06bd. Accessed 06 April 2022

Direct Benefits Of Green Economy (2020) Available from: https://www.emg-csr.com/direct-benefits-green-economy-arab-world-beyond/?msclkid=a85f3601b3fd11ec8900ddeb950b89eb. Accessed 03 April 2022

Elliott D (2022) Renewable energy and sustainable futures. Futures 32(3–4):261–274. https://doi.org/10.1016/S0016-3287(99)00096-8

Faizal M, Fong LJ, Chiam J, Amirah A (2017) Energy, economic and environmental impact of hydropower in Malaysia. Int J Adv Sci Res Manag 2(4):33–42

Fan Y, Wu SZ, Lu YT, Zhao YH (2019b) Study on the effect of the environmental protection industry and investment for the national economy: an input-output perspective. J Clean Prod 227:1093–1106. https://doi.org/10.1016/j.jclepro.2019.04.266

Fan P, Ouyang ZT, Nguyen DD, Nguyen TTH, Park H, Chen JQ (2019a) Urbanization, economic development, environmental and social changes in transitional economies: Vietnam after Doimoi. Landsc Urban Plan 187:145–155. Available from: https://viet-studies.com/kinhte/UrbanAfterDoiMoi_LUP.pdf?msclkid=c1b4d1c7b65811ec9527b0398d75469a. Accessed 05 April 2022

Hamidpour S, Fouladi N, Sedghamiz MA, Rahimpour MR (2021) Biomass technologies industrialization and environmental challenges. In: Rahimpour MR, Kamali R, Makarem MA, Manshadi MKD (eds) Advances in bioenergy and microfluidic applications. Elsevier, pp 431–453. https://doi.org/10.1016/C2019-0-03714-4

Hau E (2006) Wind turbines—fundamentals, technologies, application, economics. Springer, Berlin. https://doi.org/10.1007/3-540-29284-5

Hernandez RR, Easter SB, Murphy-Mariscal ML, Maestre FT, Tavassoli M, Allen EB, Barrows CW, Belnap J, Ochoa-Hueso R, Ravi S, Allen MF (2014) Environmental impacts of utility-scale solar energy. Renew Sustain Energy Rev 29:766–779. https://doi.org/10.1016/j.rser.2013.08.041

Khan MK, Teng JZ, Khan MI, Khan MO (2019) Impact of globalization, economic factors and energy consumption on CO_2 emissions in Pakistan. Sci Total Environ 688:424–436. https://doi.org/10.1016/j.scitotenv.2019.06.065

Khan S, Khan AA, Muhammad ASA (2022) Does emission of carbon dioxide is impacted by urbanization? An empirical study of urbanization, energy consumption, economic growth and carbon emissions—using ARDL bound testing approach. Energy Policy 164. https://doi.org/10.1016/j.enpol.2022.112908

Kieć J (2007) Renewable energy sources/Odnawialne źródła energii, Wydawnictwo MARR, Kraków

Knopper LD, Ollson CA (2011) Health effects and wind turbines: a review of the literature. Environ Health 10. Available from: https://ehjournal.biomedcentral.com/track/pdf/https://doi.org/10.1186/1476-069X-10-78.pdf. Accessed 19 April 2022

Kumar S, Madlener R (2016) CO_2 emission reduction potential assessment using renewable energy in India. Energy 97:273–282. https://doi.org/10.1016/j.energy.2015.12.131

Leung DYC, Yang Y (2012) Wind Energy development and its environmental impact: a review. Renew Sustain Energy Rev 16(1):1031–1039. https://doi.org/10.1016/j.rser.2011.09.024

Ligon FK, Dietrich WE, Trush WJJB (1995) Downstream ecological effects of dams: a geomorphic perspective. Bioscience 45:183–192

Lovich JE, Ennen JR (2011) Wildlife conservation and solar energy development in the desert southwest, United States. Bioscience 61(12):982–992. https://doi.org/10.1525/bio.2011.61.12.8

Magilligan FJ, Nislow KHJG (2005) Changes in hydrologic regime by dams. Geomorphology 71:61–78. https://doi.org/10.1016/j.geomorph.2004.08.017

Mann J, Teilmann J (2013) Environmental impact of wind energy. Environ Res Lett 8(3). https://doi.org/10.1088/1748-9326/8/3/035001

Mao G, Huang N, Chen L, Wang H (2018) Research on biomass energy and environment from the past to the future: a bibliometric analysis. Sci Total Environ 635:1081–1090. https://doi.org/10.1016/j.scitotenv.2018.04.173

Masson V, Bonhomme M, Salagnac J-L, Briottet X, Lemonsu A (2014) Solar panels reduce both global warming and urban heat island. Front Environ Sci 2. https://doi.org/10.3389/fenvs.2014.00014

Moran EF, Lopez MC, Moore N, Hyndman DW (2018) Sustainable hydropower in the 21st century. PNAS 115(47):11891–11898. https://doi.org/10.1073/pnas.1809426115

Muhammad B (2019) Energy consumption, CO_2 emissions and economic growth in developed, emerging and Middle East and North Africa countries. Energy 179:232–245. https://doi.org/10.1016/j.energy.2019.03.126

Nugent D, Sovacool BK (2014) Assessing the lifecycle greenhouse gas emissions from solar PV and wind energy: a critical meta-survey. Energy Policy 65:229–244. https://doi.org/10.1016/j.enpol.2013.10.048

Ozcan B (2013) The nexus between carbon emissions, energy consumption and economic growth in Middle East countries: a panel data analysis. Energy Policy 62:1138–1147. https://doi.org/10.1016/j.enpol.2013.07.016

Peng B, Sheng X, Wei G (2020) Does environmental protection promote economic development? From the perspective of coupling coordination between environmental protection and economic development. Environ Sci Pollut Res 27(31):39135–39148

Punch J, James R, Pabst D (2010) Wind-turbine noise: what audiologists should know. Audiol Today 8:20–31

Shahbaz M, Tiwari AK, Nasir M (2013) The effects of financial development, economic growth, coal consumption and trade openness on CO_2 emissions in South Africa. Energy Policy 61:1452–1459

Shahzad U (2020) Environmental taxes, energy consumption, and environmental quality: theoretical survey with policy implications. Environ Sci Pollut Res 27(20):24848–24862. https://doi.org/10.1007/s11356-020-08349-4

Shen N, Wang Y, Peng H, Hou Z (2020) Renewable energy green innovation, fossil energy consumption, and air pollution—spatial empirical analysis based on China. Sustainability 12(16). https://doi.org/10.3390/su12166397

Trimble JL, Van Hook RI, Fogler AG (1984) Biomass for energy: the environmental issues. Biomass 6:3–13. https://doi.org/10.1016/0144-4565(84)90003-9

Tsoutsos T, Frantzeskaki N, Gekas V (2005) Environmental impacts from solar energy technologies. Energy Policy 33(3):289–296. https://doi.org/10.1016/S0301-4215(03)00241-6

Wang SS, Zhou DQ, Zhou P, Wang QW (2011) CO_2 emissions, energy consumption and economic growth in China: a panel data analysis. Energy Policy 39:4870–4875. https://doi.org/10.1016/j.enpol.2011.06.032

Wang C, Prinn R (2010) Potential climatic impacts and reliability of very large-scale wind farms. Atmos Chem Phys 10(4). https://doi.org/10.5194/acp-10-2053-2010

Wang C, Prinn RG (2011) Potential climatic impacts and reliability of large-scale offshore wind farms. Environ Res Lett 6(2). http://doi.org/https://doi.org/10.1088/1748-9326/6/2/025101

Wu Y, Zhao F, Liu S, Wang L, Qiu L, Alexandrov G, Jothiprakash V (2018) Bioenergy production and environmental impacts. Geosci Lett 5(14). https://doi.org/10.1186/s40562-018-0114-y

Yang X, Li N, Mu H, Ahmad M, Meng X (2022) Population aging, renewable energy budgets and environmental sustainability: does health expenditures matter? Gondwana Res 106:303–314. https://doi.org/10.1016/j.gr.2022.02.003

Yu L, Jia B, Wu S, Wu X, Xu P, Dai J, Wang F, Ma L (2019) Cumulative environmental effects of hydropower stations based on the water footprint method—Yalong River Basin, China. Sustainability 11(21):5958. https://doi.org/10.3390/su11215958

Ziębik A, Szorgut J (1997) Basics of energy economy/Podstawy gospodarki energetycznej. Wydawnictwo Politechniki Śląskiej, Gliwice

Zoundi Z (2017) CO_2 emissions, renewable energy and the environmental Kuznets curve, a panel cointegration approach. Renew Sustain Energy Rev 72:1067–1075. https://doi.org/10.1016/j.rser.2016.10.018

Selected Social Aspects of the Green Transformation

Anna Barwińska-Małajowicz and Miroslava Knapková

Abstract The green transformation is a long-lasting process with many direct and indirect impacts and consequences, not only on countries and their economies, but also on households, individuals, and their well-being. These relations are manifested in various areas, which are elaborated at different levels both in theoretical and practical terms. The chapter addresses the social conditions of the green transformation. Attention is focused on the selected determinants of green growth, including social well-being, the socio-economic context considered in the sets of green economy indicators which are developed by the OECD, and the green behavior of households. The desk research analysis based on the available literature, content analysis of the documents, and secondary statistical data processing were used to elaborate the study. Although green transformation is primarily concerned with economic development and the green economy in general, without accepting the requirement of the social aspect, green transformation cannot be considered successful and complete.

1 Introduction

In many countries, changes in investment models, technology, production and consumption patterns related to sustainable development and aimed at shifting from high to low-carbon systems are observed. An important task of the climate-ecological and climate-energy transition is the implementation and use of technological solutions that are environmentally friendly. Governments and societies are confronted with the need to understand the climate and environmental challenges in a civilizational context and in terms of new rules of behavior associated with the green transformation.

A. Barwińska-Małajowicz (✉)
Institute of Economics and Finance, University of Rzeszów, Rzeszów, Poland
e-mail: abarwinska@ur.edu.pl

M. Knapková
Faculty of Economics, Matej Bel University in Banská Bystrica, Banská Bystrica, Slovakia
e-mail: miroslava.knapkova@umb.sk

The starting point for a transition towards a green economy is not the same for all countries, sectors and households. Not all are equally capable of meeting the challenges of this transition. Countries and regions with a higher share of fossil fuels in the energy mix, higher GHG emissions, higher energy intensity and lower GDP per capita face far greater challenges. High CO_2 emitting sectors, and regions where a large part of the economy depends on such sectors, face the most difficult transition tasks. New skills, public awareness of green technologies and changes in the mindset of many different actors will be needed.

For different social groups, countries and regions, there are not only significant differences in the assessment of the opportunities, costs and benefits of moving to a green economy, but also divergent views on approaches to achieving the social, environmental and economic objectives associated with the concept of sustainable development. In this context, the transition to a green economy with social justice and poverty reduction is becoming one of the important challenges.

The Rio + 20 process highlighted this explicit link between the green economy and the goals of sustainable development and poverty eradication, emphasizing the importance of the social dimension of development. In turn, in January 2018, the World Economic Forum in Davos proposed an alternative system for assessing economic development based on the Inclusive Development Index (IDI), which reflects, among other things, the social component of the economy (The Inclusive Development Index ..., http).

The aim of the chapter is to present selected social aspects of green economy development. Due to the volume limitations of the publication, the focus is on selected determinants of green growth, including social well-being, the socio-economic context considered in the sets of green economy indicators developed by the OECD and green household behavior.

2 Theoretical Conceptual Frameworks

Many approaches and concepts emerge in the international debate on the development of the green economy, which analyze its different contexts, including the social aspect. This chapter refers to selected contributions from the literature related to this issue.

According to the United Nations Environment Programme, a green economy is characterized by improving human well-being, quality of life and social equity while reducing environmental risks, ecological resource scarcity, carbon emissions and increasing resource productivity (Kothari et al. 2014).

The social aspect of green economy development is a new perception of the world, a new social approach to the environment with rational use of natural resources, with a responsible approach to nature. The green economy is a kind of new philosophy, which aims to overcome, reduce and compensate for the negative consequences of the functioning of the market economy, which adversely affected the state of the environment surrounding humans (Stukalo and Simakhova 2019, p. 91).

There is a need for coordinated efforts by all governments to establish a detailed set of feasible and effective policies for the green economy, as Jackson (2011) pointed out. He also highlights actions that governments can take to stimulate the transition to a green economy. He outlines three main areas of action, among which he identifies changing the social logic of consumerism. Within this change, he emphasizes the following elements: "sharing the work and improving the work-life balance, tackling systemic inequality, measuring prosperity, strengthening human and social capital, reversing the culture of consumerism" (Jackson 2011).

Jacob et al. (2015) addresses the issue of the impact of the green economy on employment. He pointed out that policy instruments for the transition to a green economy cannot be directly transferred from one country to another, but must be adapted to the specific economic, institutional and social context of the green economy. He pointed out that all approaches have in common the perception of the green economy as a way to reconcile the three pillars: economic, environmental, social (Jacob et al. 2015, p. 19; World Bank 2012, p. 24). In this context, employment aspects are addressed under two distinct themes:

(1) inclusiveness—economic growth and employment in the green economy should bring benefits to all groups in society (access to the labour market, education and employment opportunities, the need for socio-political intervention targeting the poorest members of society),
(2) decent work—contemporary development challenges: "achieving environmental sustainability and turning the vision of decent work for all into a reality" (Jacob et al. 2015, p. 17); green jobs should also be considered as a decent job (UNEP/ILO/IOE/ITUC 2008; ILO 2012).

The social thread is also raised by Barbier (2012), who takes the attitude that the green economy aims to combine environmental protection and the fight against poverty.

In the other hand, D'Amato et al. (2017) consider and analyze the concepts of the circular economy, the green economy and the bio economy, which—despite their different assumptions and operationalization strategies—share a common denominator in terms of reconciling economic, environmental and social objectives. They include the following areas among the main aspects covered by the social dimension of sustainability: education and training, social justice, democracy, health, quality of life and well-being, social inclusion, social capital, social networks, security, employment and income equality, governance and social cohesion, cultural traditions, recreation and tourism (D'Amato et al. 2017).

Mazilu (2013) focuses on the issue of sustainable tourism in the era of the green economy. He points out that the tourism sector is extremely important from a global resource perspective, and therefore even small changes towards greening can make a big difference. He points out that tourism has been identified by organizations such as the World Bank, IMF and UN as a key economic sector in the global transition from a brown to a green economic system, and stresses the need for the tourism industry and host countries to take decisive action to reduce the negative environmental impact of foreign travel and to reduce the negative social impacts associated with the existence of foreign tourism enclaves in developing countries.

The connection between the concept of green economy and the idea of social and environmental justice is also pointed out by Ehresman and Okereke (2015). They propose a systematization of possible interpretations of the green economy, which is reflected by a synthesis of existing typologies of environmental justice.

The way in which the three pillars of sustainable development (economic growth, social justice and environmental protection), already mentioned, influenced each other between 1987 and 2013 was studied by Huh and Kim (2021). Using twenty-six OECD countries as an example, they examined the correlation of nine representative variables in these three categories.

The social aspect also runs through the reflections of many other authors and documents of various international institutions and organizations, such as Hussien et al. (2016), Pop et al. (2011), Kantola et al. (2017), UNRISD (2012), UNRISD/RPB 12 (2012), Musyoki (2012), Cook et al. (2012), Hiraldo and Tanner (2011). Most authors, however, emphasize more often the economic component of green growth.

3 Social Aspect of the Green Economy as Seen by the United Nations

The definitions of green economy available in the literature emphasize respect for the environment and include environmental costs in economic activities. Among the definitional approaches, not only the economic and ecological aspect (expressed mainly in terms of reduction of CO_2 emissions, and resource efficiency) is clear, but also the social aspect, which in general consists of counteracting social exclusion (Towards a Green Economy..., http, pp. 2–4; Ryszawska 2013, p. 28).

The United Nations (UN), explicitly highlighting the social aspect of the green economy, emphasizes the importance of the so-called Inclusive Green Economy (IGE), i.e. one that improves human well-being and builds social equality while reducing environmental risks and resource consumption (Sidorczuk-Pietraszko 2018, p. 74).

At the level of international institutions, sets of green economy indicators have been identified, among which the best known is the set developed by the OECD. This set considers not only the relationship between the economy and the environment, but also indicates social aspects directly related to the economy or the environment. The key areas for which indicators have been proposed are defined as follows (OECD 2011):

(1) natural asset base,
(2) environmental and resource productivity,
(3) environmental quality of life,
(4) economic opportunities and policy responses,
(5) the socio-economic context.

Table 1 The socio-economic context of the green economy (own elaboration based on OECD 2011)

Main indicator groups	Topics covered	Related OECD works
Economic growth, productivity and competitiveness	Economic growth and structure Productivity and trade Inflation and commodity prices	Economic outlook, Economic surveys Going for growth
Labor markets, education and income	Labor markets (employment/unemployment) Socio-demographic patterns Income and education	National accounts, Productivity database Employment outlook Education at a glance Health at a glance Society at a glance

Table 1 presents the last-mentioned area, i.e. the socio-economic context (indicating main indicator groups, topics covered, related OECD work).

The OECD, as part of its statistical data collection, maintains a database of green economy indicators for member countries, candidate countries, major partners and other selected countries. The OECD attempts to monitor progress in the transition to a green economy, in cooperation with other international organizations such as UNEP, the European Commission and various international institutes. The indicators needed to measure this progress are based on existing OECD work that is being aligned with the Green Growth Strategy. In the socio-economic context, OECD work mainly addresses two areas (OECD 2011):

(1) measuring sustainable development—the OECD is promoting the development of indicators and consistent measurement approaches. Work focuses on improving the measurement of different types of capital, with an emphasis on human and social capital;
(2) measuring well-being—the OECD is promoting the development of better measures and indicators of well-being and social progress (with an emphasis on well-being and sustainability) to be used alongside standard economic measures such as gross domestic product.

Within the social dimension, the following indicators are included in OECD databases (OECD Stat, http):

(1) population, ages 0–14, % total,
(2) population, ages 15–64, % total,
(3) population, ages 65 and above, % total,
(4) women, % total population,
(5) total fertility rate, children per woman,
(6) life expectancy at birth,
(7) net migration,
(8) population density, inhabitants per km^2.

The presented green economy measurement context refers to the national dimension (data are collected at the national level) (Wyszkowska and Rogalewska 2014;

Sidorczuk-Pietraszko 2018). In contrast, proprietary indicator sets are usually used at the regional level, depending on the availability of data for the regional level. For example, Li and Lin (2016) proposed indicators to assess China's green economy performance and green productivity growth, in which they considered economic expansion, resource conservation and environmental protection by studying 275 cities in China from 2003 to 2012.

Most studies on the green economy focus on its selected aspects, such as CO_2 emissions, renewable energy and green jobs (Wei et al. 2012). In contrast, Shi et al. (2016) attempted to assess the development of the green economy in China by studying 15 sub-regional cities. Among the evaluation criteria, they also tried to capture the social element i.e. social livelihood of the people.

4 Selected Conditions and Determinants of Green Transformation

In 2012, the G20 introduced the concept of inclusive green growth, recognizing it as a cross-cutting priority of the G20 development agenda. The Seoul meeting (19–20 March 2012) emphasized the need to focus efforts on inclusive green growth, pointing out that it is a key element in achieving long-term sustainable development (OECD 2015).

It is important to stress that there is no universal and uniform procedure for implementing green growth. The ways of implementing a green transition depend on political and institutional conditions, the level of economic development, the resources available, as well as on social conditions. Highly developed countries, emerging economies and developing countries face diverse challenges (OECD 2011, pp. 10 and 23).

Green growth scenarios for individual countries should be based on regional or local conditions, adaptive capacity and the priorities formulated by individual stakeholders. Priorities for action for developing country governments include ensuring food security, basic education and the supply of basic services (e.g. water supply, sanitation systems). In addition, the economies of these countries are more dependent on natural resources and highly exposed to the effects of climate change, and therefore adaptation to the adverse effects of climate change and proper management of natural resources are crucial for the further development of these countries (Kasztelan 2015). At the same time, international cooperation is not insignificant for ensuring the effectiveness of actions implemented in this area (OECD 2010, pp. 13–14).

For developed countries, one of the important tools that can be used in the introduction of the green economy is the digital economy, which can promote the accelerated development of the green economy. For example, digital technology can be used in the consumer market to develop green products and create a platform for green consumption. Digital economy researchers can also use interdisciplinary frameworks to explore sustainable energy issues. As the Declaration on the Digital

Economy: Innovation, Growth and Social Prosperity (Cancún Declaration) highlights, the digital economy is a huge catalyst not only for innovation and growth, but also for social prosperity (OECD 2016b). The demands in the aforementioned declaration are to promote "a more sustainable and inclusive growth focused on wellbeing and equality of opportunities, where people are empowered with education, skills and values, and enjoy trust and confidence" (OECD 2016b).

In turn, the Declaration on Enhancing Productivity for Inclusive Growth (OECD 2016a), while pointing out that productivity is a multidimensional concept, indicates that a more inclusive approach to productivity growth should be the key to enhancing productivity growth for the benefit of all segments of society. This approach should be based, on the one hand, on investments in people (education, skills, health) and in quality jobs, and, on the other hand, it should support the development of lagging regions and firms to exploit their productive potential, while promoting competition, especially in new technology sectors. The Declaration proposes ways to help countries understand their productivity challenges while promoting inclusive growth.

Despite the pandemic crisis, as well as many other difficulties, conflicts and political and socio-economic problems, the governments of many countries are adopting systematic steps aimed at intensifying climate and environmental action. For example, on 31 March 2022, government representatives from OECD countries, the European Union, several non-member countries (Argentina, Brazil, Bulgaria, Croatia, Egypt, Indonesia, Kazakhstan, Peru and Romania), and representatives of International Organizations have committed in a formal OECD Declaration to intensify work on climate and environment. Among the actions pledged—in addition to those to reduce biodiversity loss, plastic pollution, or aligning finance with environmental goals—attention was also paid to the social dimension of the green economy, pointing to two key needs:

(1) the need for greater collaboration not only with NGOs, civil society and the private sector, but also with women, youth, indigenous peoples, vulnerable and underserved communities and regions; collaboration oriented towards supporting policies for sustainable development and net zero greenhouse gas emissions and positive transformation of nature;
(2) deepening "analysis of the distributional effects of national environmental policies, considering social and gender dimensions, and national approaches to environmental justice, with a goal of improving equality and equity, fairness, inclusiveness, citizens' awareness and their meaningful engagement" (OECD 2022).

The OECD Declaration underscores the importance of the work undertaken by the OECD and EPOC to help shape transformative and innovative policies and standards in such a way that their implementation can lead to building stronger, sustainable and inclusive economies (OECD 2022).

It is difficult to clearly divide the determinants of economic development into economic and non-economic, as development is the result of the interaction of both groups of factors. Economic factors are very important, but non-economic factors are equally important, especially social determinants, which are more dynamic and stronger. The literature gives different classifications of factors of a social kind that influence economic development (Orłowska 2018, p. 48), playing an important role in shaping its pace and directions. Researchers in the social sciences have realized this relatively recently. The social determinants of economic development include, for example, family, class structure, origin, religion, traditions, attitudes, beliefs and culture, etc. The tempo, trends and nature of economic development are also influenced by factors such as a society's adaptability, its attitude (openness) to innovation and change, social mobility, social well-being, the attitude of governments and societies to economic, social and political change (The Social Determinants..., http). Some of the above-mentioned determinants have a stronger influence on green growth, others play a smaller role in this respect. The progress of the green economy process should be made considering country-specific social determinants. Successful transition from an economy based on carbon-intensive energy sources to a green economy requires that attention is paid to ensuring that the process leads to win–win changes. What is needed are people-centered solutions, and measures that consider people's needs and concerns, and that also address the issue of social well-being mentioned earlier.

4.1 Social Well-Being and Employment

Social well-being consists of social security, family life, social relationships (private success) and achievement and respectability (public success) (Mirski 2009, p. 172). In the index dimension, one can try to capture one of the key components of social well-being, which is social security. It can be considered in two contexts:

(1) material—providing society with the means to support oneself and one's family (work, social benefits) and social infrastructure facilities (e.g. educational institutions, health care facilities, cultural institutions),
(2) social—ensuring order in social relations (What is social security?, http).

The material aspect of social security is, for most of society, closely related to employment, which in the context of green growth is an extremely important development factor. In this context, the division of employment into "green", "less green" and "not green" will help to better understand the impact of this factor on the greening of the economy. Taking carbon production and the intended CO_2 emissions target as a criterion, employment can be divided into four main categories, which are shown in Fig. 1.

Categories I and II are strictly green jobs that exist in low-carbon (sustainable) production, with Category I jobs being the greenest (both the environmental impact of production and the nature of the activity are geared towards reducing CO_2 emissions). The greening of Category II jobs, on the other hand, is due to the fact that they are not associated with any negative long-term effects on the global climate. Categories III and IV, on the other hand, are associated with carbon-intensive production (characteristic of the so-called "brown" sectors), with jobs in Category III being

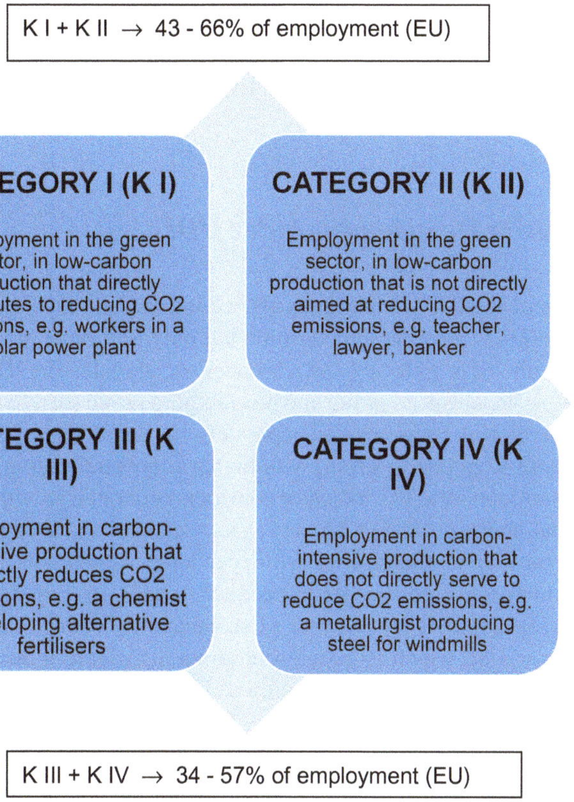

```
                K I + K II  →  43 - 66% of employment (EU)
```

CATEGORY I (K I)

Employment in the green sector, in low-carbon production that directly contributes to reducing CO_2 emissions, e.g. workers in a solar power plant

CATEGORY II (K II)

Employment in the green sector, in low-carbon production that is not directly aimed at reducing CO_2 emissions, e.g. teacher, lawyer, banker

CATEGORY III (K III)

Employment in carbon-intensive production that directly reduces CO_2 emissions, e.g. a chemist developing alternative fertilisers

CATEGORY IV (K IV)

Employment in carbon-intensive production that does not directly serve to reduce CO_2 emissions, e.g. a metallurgist producing steel for windmills

```
                K III + K IV  →  34 - 57% of employment (EU)
```

Fig. 1 Categories of employment by type of production in the context of the green economy (own elaboration based on Towards a greener economy: the social dimensions, 2011. International Labor Organization, Geneva)

associated with activities that aim to reduce CO_2 emissions, and can therefore be included in the broader definition of green jobs. In the longer term, however, it is desirable to shift jobs from Category III to Category I (Towards a greener economy... 2011, pp. 14–15).

The size of each employment category is important in determining the potential effects of the introduction of the green economy. However, detailed employment data by the categories outlined above is not easily available. Estimates for the European Union indicate that Categories I and III (industries focused on reducing CO_2 emissions) account for a small percentage of total employment. Most jobs are in high-carbon-intensive sectors (HCIS), and above all in Category IV, while the share of emitting industries in GDP is relatively high (Towards a greener economy... 2011, p. 16–17).

It is worth noting here that according to Pai et al. (2021), more than 80% of jobs in the energy sector are expected to be related to renewable energy sources by 2050, with the largest job gains in the renewable energy sector coming from solar and wind power.

4.2 Inclusive Development Index (IDI)

The analysis of the IDI (which ranks 103 countries for which the required data are available) makes it possible to determine not only economic growth itself, but also to assess whether it is noticeable on a broader social scale. The IDI ranks economies in two groups: an advanced group and an emerging group (results are collected separately for developed countries and separately for developing countries). Countries receive a score from one to seven, with higher scores translating into greater inclusion in a given economy. The indicator considers economic growth and development, inclusion and intergenerational balance. The social context is captured in the IDI by measuring poverty or the size of the dependency burden (the number of people of non-productive age per people of productive age). Among developed countries, the best performing country in terms of sustainability of economic growth in 2018 was Norway (6.08), which is mainly due to its high score in terms of intergenerational balance. It was immediately followed by two countries ex aequo: Iceland and Luxembourg (6.07) (Fig. 2).

Among developed countries, relatively small European countries are the most inclusive. Among the world's largest economies (G7), Germany ranks best (12th place in the ranking).

However, among emerging countries, Lithuania was ranked first. Hungary came second, followed by Azerbaijan and Latvia, and Poland in fifth place (Fig. 3).

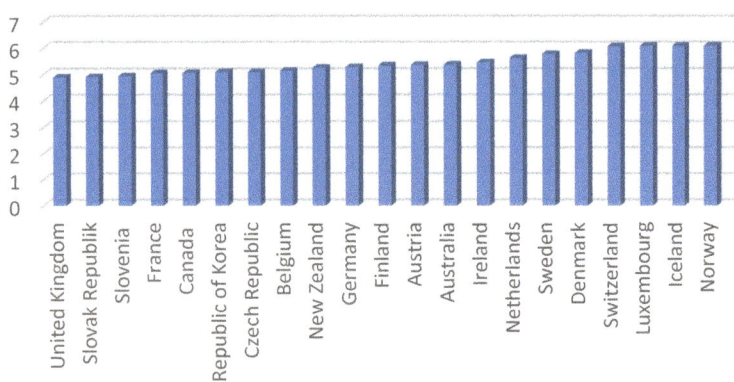

Fig. 2 Leading advanced economies according to the Inclusive Development Index in 2018 (own elaboration based on The Inclusive Development Index 2018. Summary and Data Highlights, World Economic Forum)

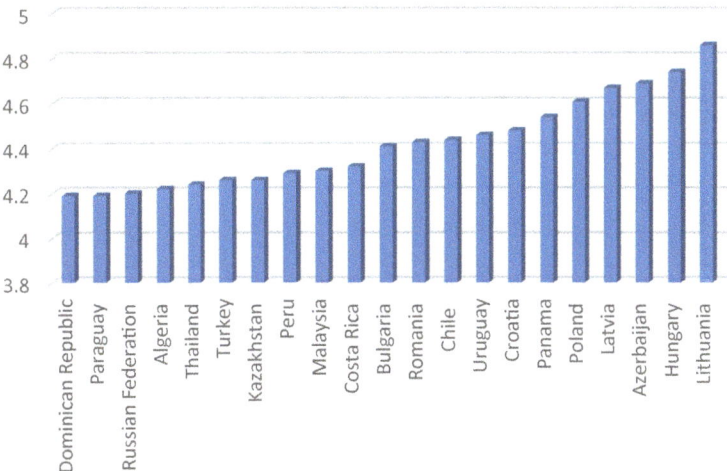

Fig. 3 Leading emerging economies according to the Inclusive Development Index in 2018 (own elaboration based on The Inclusive Development Index 2018. Summary and Data Highlights, World Economic Forum)

In this part of the ranking, European economies have topped the ranking, thanks primarily to their high GDP growth rates in recent years. Furthermore, the IDI analysis indicates that inclusive economic progress is correlated with higher levels of interpersonal trust. This points to the need for a more human-centered approach to improving cohesion in increasingly divided societies. Nevertheless, some Asian economies show that it is possible to maintain high levels of trust despite only average levels of inclusive development (The Inclusive Development Index… http).

4.3 Households' Environmental Behavior

The green transformation is, on the one hand, a huge opportunity for industry, for example by creating new markets for new green technologies and products. On the other hand, it poses many challenges for individual economic entities, particularly in relation to reducing greenhouse gas emissions. Reducing the carbon intensity of the economy focuses, inter alia, on such areas as the energy system, construction, industry, transport and households. It is worth taking a closer look at the latter area, since the functioning of households is closely linked to the social dimension of the management process. The social process of management, serving to shape the sense of well-being of the whole society, includes all members of households, whose behavior has specific consequences for the green transformation.

The process of greening households includes activities of various types, among which the following can be mentioned as examples (Garbage Segregation… http):

1. ecological separation and management of rubbish,
2. thermo-modernization, i.e. proper insulation of one's own house,
3. verifying actual demand and reducing electricity consumption, as well as investing in renewable energy sources e.g. photovoltaics (according to research, solar photovoltaics are among the cheapest forms of electricity production in many countries and market segments (Vartiainen et al. 2019, p. 439).

One of the fundamental problems of modern ecology is the excessive production of rubbish. The share of rubbish produced by households in the total waste produced by all entities is not as high as in the case of the post-production waste sector (the leader of which is the mining and quarrying industry). For example, in Poland it accounts for less than 10% of all waste produced annually in the country (data of the Central Statistical Office (CSO) from 2018). However, in terms of shaping general social awareness, education concerning both waste separation and recycling is of great importance, precisely at the level of individual households.

Pro-environmental attitudes and behaviors of households depend on the characteristics specific to a particular household and its income. Many categories of households, e.g. families with many children, families with disabled or sick people, families with unemployed people, socially excluded people, people without access to public transport, may suffer from increased costs of living resulting from the need to adapt to the requirements of the green transformation. Effective social policies must protect them from price increases. On the other hand, for households with, for example, their own real estate (house, flat), investments in improving energy infrastructure are beneficial as they increase the value of the asset and reduce maintenance costs.

5 Summary

The Green Transformation will only be successful if businesses, governments and individual citizens work together to achieve global decarbonization through investment, taxation, subsidies and behavioral change. In the social dimension, it is behavioral change that is crucial, starting with individuals, households, businesses, national governments and societies.

It is a great challenge to take care of such a direction of socio-economic development, in which the challenges of green transformation would be perceived not only as threats or forced adaptations, but also as development opportunities. The extent of action taken in this regard will vary within countries. Some will continue to focus on economic development, but with a greater emphasis on environmental protection. A small group of countries will see the so-called 'green' sectors of the economy as a potential catalyst for the creation of new jobs or economic activity. For others, the green growth revolution may only be a short-term plan to stimulate economic growth. In some countries, long-term structural changes will be initiated, with green growth treated as an instrument to support a long-term vision of development.

References

Barbier E (2012) The green economy post Rio+20. Science 338:887–888
Co to jest bezpieczestwo socjalne? (What is social security?). https://businessinsider.com.pl. Accessed 15 April 2022
Cook S, Smith K, Utting P (2012) Social dimensions of green economy transition. No. 10 (forthcoming)
D'Amato D, Droste N, Allen B et al (2017) Green, circular, bio economy: a comparative analysis of sustainability concepts. J Cleaner Prod 168:716–734. https://doi.org/10.1016/j.jclepro.2017.09.053
Ehresman T, Okereke C (2015) Environmental justice and conceptions of the green economy. Int Environ Agreements Polit Law Econ
Hiraldo R, Tanner T (2011) The global political economy of REDD+: engaging social dimensions in the emerging green economy. Social dimensions of green economy and sustainable development, Occasional Paper No. 4. UNRISD, Geneva
Huh T, Kim YY (2021) Triangular trajectory of sustainable development: panel analysis of the OECD countries. Int J Environ Res Public Health 18:2374. https://doi.org/10.3390/ijerph18052374
Hussien ME, Alam RZ, Siwar C et al (2016) Green economy models and energy policies towards sustainable development in Malaysia: a review. Int J Green Eco 10:89–106
ILO (2012) Working towards sustainable development: opportunities for decent work and social inclusion in a green economy. International Labor Office, Geneva
Jackson T (2011) Philosophical and social transformations necessary for the green economy. Background Paper for the World Economic and Social Survey
Jacob K, Quitzow R, Bär H (2015) Green jobs: impacts of a green economy on employment. Deutsche Gesellschaft für Internationale Zusammenarbeit (GIZ) GmbH
Kantola J, Liu Y, Peura P et al (2017) Innovative products and services for sustainable societal development: current reality, future potential and challenges. J Cleaner Prod 162:S1–S10. https://doi.org/10.1016/j.jclepro.2017.07.091
Kasztelan A (2015) Zielony wzrost jako nowy kierunek rozwoju gospodarki w warunkach zagrożeń ekologicznych (green growth as a new direction for the economy under the environmental threats). St Ekonom 2:185–208
Kothari A, Demaria F, Acosta A (2014) Buen Vivir, degrowth and ecological Swaraj: alternatives to sustainable development and the green economy. Development 57(3–4):362–375. https://doi.org/10.1057/dev.2015.24
Li J, Lin B (2016) Green economy performance and green productivity growth in China's cities: measures and policy implication. Sustainability 8:947. https://doi.org/10.3390/su8090947
Mazilu M (2013) Green tourism in the age of green economy. Int J Econ Stat 3(1):140–147
Mirski A (2009) Dobrostan jako kategoria społeczna i ekonomiczna (well-being as a social and economic category). Panstwo i Spoleczenstwo (2)
Musyoki A (2012) Emerging policy for a green economy and social development in Limpopo, South Africa. No. 8
OECD (2010) Interim report of the green growth strategy: implementing our commitment for a sustainable future. In: Meeting of the OECD Council at Ministerial Level 27–28 May 2010, C/MIN(2010)5
OECD (2011) Towards green growth: monitoring progress. OECD Indicators
OECD (2015) A toolkit of policy options to support inclusive green growth. In: Submission to the G20 Development Working Group by the AfDB, the OECD, the UN and the World Bank. http://www.oecd.org/greengrowth/toolkit.pdf. Accessed 22 Mar 2022
OECD (2016a) Declaration on enhancing productivity for inclusive growth. OECD/LEGAL/0425
OECD (2016b) Declaration on the digital economy: innovation, growth and social prosperity (Cancún declaration), OECD/LEGAL/0426
OECD (2022) Declaration on a resilient and healthy environment for all. OECD/LEGAL/0468

OECD Stat. https://stats.oecd.org/Index.aspx?DataSetCode=GREEN_GROWTH. Accessed 12 April 2022

Orlowska J (2018) Czynniki i bariery rozwoju lokalnego (factors and barriers of local development). In: Gruszewska E, Karpilska K, Protasiewicz A (eds) Wspólczesne problemy ekonomiczne w badaniach mlodych naukowców, t. 1. Wzrost, rozwój i polityka gospodarcza. Polskie Towarzystwo Ekonomiczne, Warszawa

Pai S, Emmerling J, Drouet L et al (2021) Meeting well-below 2 C target would increase energy sector jobs globally. One Earth 4. Elsevier Inc.

Pop O, Dina GC, Martin C (2011) Promoting the corporate social responsibility for a green economy and innovative jobs. Procedia-Soc Beh Sci 15:1020–1023. https://doi.org/10.1016/j.sbspro.2011.03.232

Ryszawska B (2013) Zielona gospodarka w dokumentach strategicznych Unii Europejskiej (green economy in the strategic documents of the European Union). Ekonomia i Środowisko 3(46)

Segregacja smieci (2022) inwestycja w OZE i termomodernizacja, czyli bycie "eko" we wlasnym domu w 3 krótkich punktach (separation of rubbish, investment in RES and thermo-modernisation, i.e. being "eco" in your own home in 3 short points). https://pvge.pl. Accessed 21 Mar 2022

Shi B, Yang H, Wang J et al (2016) City green economy evaluation: empirical evidence from 15 sub-provincial cities in China. Sustainability 8(551). https://doi.org/10.3390/su8060551

Sidorczuk-Pietraszko E (2018) Zielona gospodarka w polityce rozwoju województwa podlaskiego - strategia i postupy (green economy in the development policy of Podlaskie voivodeship - strategy and steps forward). Zeszyty Naukowe WSE 1(2):72–91

Stukalo N, Simakhova A (2019) Social dimensions of green economy. Filosofija, Sociologija 30(2)

The Inclusive Development Index (2018) Summary and data highlights. World Economic Forum. https://www3.weforum.org/docs/WEF_Forum_IncGrwth_2018.pdf. Accessed 21 Feb 2022

The Social Determinants and Consequences of Economic Development (2022). https://www.you rarticlelibrary.com. Accessed 21 Mar 2022

Towards a green economy: pathways to sustainable development and poverty eradication—a synthesis for policy makers. www.unep.org. Accessed 12 April 2022

Towards a greener economy: the social dimensions (2011) International Labor Organization, Geneva

UNEP/ILO/IOE/ITUC (2008) Green jobs: towards decent work in a sustainable, low-carbon world

UNRISD (2012) From green economy to green society: bringing the social to Rio+20, Geneva

UNRISD/RPB 12 (2012) social dimensions of green economy, Geneva

Vartiainen E, Masson G, Breyer C et al (2019) Impact of weighted average cost of capital, capital expenditure, and other parameters on future utility-scale PV levelised cost of electricity. Progress in Photovoltaics: Research and Applications. Wiley

Wei C, Ni J, Du L (2012) Regional allocation of carbon dioxide abatement in China. China Econ Rev 23:552–565

World Bank (2012) Inclusive green growth. World Bank, Washington DC. http://siteresources.wor ldbank.org. Accessed 21 Feb 2022

Wyszkowska D, Rogalewska A (2014) Monitorowanie zielonej gospodarki w ujeciu organizacji miedzynarodowych (International organisations' monitoring of the green economy). Optimum. St Ekonom 3(69)

The Directions of Financing the Green Energy Transformation

Anna Spoz and Magdalena Ziolo

Abstract Financing the green transformation constitutes another vital area for countries around the world. Many governments are allocating ever-increasing funds for financing green technologies and the decarbonisation of their economies. By 2050 the European Union is supposed to become the world's first bloc of climate-neutral countries. This requires substantial investment on the part of the EU and the domestic public sector, as well as the private sector. The investment plan for the European Green Deal presented last year aims at a significant increase in the activity of EU countries in this area. The chapter presents the most relevant possibilities of financing the green transformation, also providing examples of similar regulations introduced in EU member states.

1 Introduction

Climate risk, and more broadly, environmental risk is one of the key types of risk, the impact of which affects both the financial and real spheres (Global Risks Report 2021). This impact is reflected in the impact on the financial results of all market participants and, above all, on the quality of life. The effects of climate change and forecasts of the directions of climate risk impact will trigger a number of new phenomena that should be effectively counteracted, inter alia, climate migrations, diseases determined by smog, changes in the labor market. As a consequence, a number of countries and international organizations prioritized preventive and mitigating actions against the effects of environmental risk in their strategies. A key element of the impact is the reduction of greenhouse gas emissions, which are the most responsible for climate change. Air quality and responsibility for its pollution

---

A. Spoz (✉)
Institute of Economics and Finance, The John Paul II Catholic University of Lublin, Lublin, Poland
e-mail: aspoz@kul.lublin.pl

M. Ziolo
Faculty of Economics Finance and Management, University of Szczecin, Szczecin, Poland
e-mail: magdalena.ziolo@usz.edu.pl

© The Author(s), under exclusive license to Springer Nature Switzerland AG 2023
I. Bąk and K. Cheba (eds.), *Green Energy*, Green Energy and Technology,
https://doi.org/10.1007/978-3-031-12531-7_6

is a common matter, and air is a public good, therefore it cannot be effectively influenced by its quality without the cooperation of all countries on a global scale, hence the actions for the climate and the environment are expressed in the form of the so-called climate summits (COP, the last COP26 was held in 2021 in Glasgow), at the level of international organizations and remedial programs developed by these organizations, including programs of the United Nations (The United Nations Environment Program (UNEP), the European Green Deal, Fit for 55, or at the level of individual governments and local governments.

Preventive actions and reducing the negative impact of environmental risk have a complex dimension and relate to transformations both within the financial sphere (including tax systems, external financing) and the real sphere, e.g., thermomodernization, decarbonization. All sectors of the economy are and will be subject to change. First of all, attention is drawn to the necessity of the widespread introduction of energy from renewable sources, the transport sector must be completely transformed towards low-emission, buildings and structures, especially public ones (schools, hospitals, offices), require urgent thermal modernization, which also applies to households. These are only selected directions of changes that require significant financial outlays. One of the most advanced actions to reverse negative environmental changes is currently the European Union, which has implemented or is implementing a number of solutions supporting this process, including The EU taxonomy (Regulation (EU) 2020/852 ('Taxonomy Regulation')). The energy transformation is not possible without the provision of funds for its financing. The scale of expenditures necessary to finance planned activities requires support from both public funds and public–private financing mechanisms, and in turn private investment must be stimulated with the use of fiscal policy tools.

The aim of the chapter is to discuss the mechanism of financing the energy transformation, taking into account public, private and public–private financing tools, with an indication of the existing regulations supporting this process. In particular, the following research questions correspond with the aim of this chapter: (1) How should the public financial system be designed and what solutions should be introduced in the field of public finances in order to effectively support the energy transformation and the accompanying challenges? (2) What is the role of the financial sector and enterprises in supporting the energy transition? (3) How should the mechanisms of cross-sectoral cooperation, including those based on public–private partnership, be used to effectively carry out the energy transformation?

The remainder of this chapter is organized as follows. Section 2 briefly reviews the related literature in the scope of energy transformation. Section 3 describes the directions and the mechanisms of financing the green energy transformation, Sect. 4 presents case studies and issues related to the implementation of financing the green energy transformation, and Sect. 5 concludes the chapter.

2 Literature Review

The growing awareness of human impact on the environment and the irreversibility of the climate change as well as and the responsibility of current societies for the living conditions of future generations made the issue of green economy the subject of many scientific studies. The first to use the term "green economy" were Pearce et al. (1989). In the book "Blueprint for a green economy", the authors expressed their belief that the environment and the economy should cooperate with each other. However, an attempt to conceptualize the idea was made two years later by Jacobs (1991). For the next several years, this concept did not gain much interest from scientists. This was partly due to the dynamically developing concept of sustainable development (Brown et al. 2014). The real development of the green economy took place after 2008, when this concept started to be seen not only as a possible response to the global financial crisis, but also to growing environmental problems (Bina and La Camera 2011). In 2008, UNEP established the Green Economy Initiative and a year later the Global Green New Deal (Merino-Saum et al. 2019). Currently, the concept of green economy is the subject of research in the context of its relationship with sustainable development (Lavrinenko et al. 2019; Aldieri and Vinci 2018; Mikhno et al. 2021), green growth (Hickel and Kallis 2020; Stoknes and Rockström 2018), green finance (Zhang et al. 2021), technologies and innovations, that is the transition to a low-emission and ultimately zero-emission economy (Yumei et al. 2022; Kasayanond 2019; Ying et al. 2021; Zhao et al. 2019; Conti et al. 2018; Liu and Dong 2021).

Resulting from the Paris Agreement (2015) obligations of states to undertake actions aimed at limiting the average temperature increase on Earth and to include the development of renewable energy in the energy strategy of countries that respect the provisions of the agreement became a priority (Koval et al. 2021). Independent studies by Chu and Majumdar (2012) and Zhou et al. (2015) shows that government policy has a large impact on the development of renewable energy sources. It directly or indirectly reduces the costs related to renewable energy sources, thus increasing its competitiveness against traditional energy (Donastorg et al. 2017). The state can stimulate the development of renewable energy using financial incentives (i.e., tax incentives, loan, feed-in-tariff) and Renewable Portfolio Standard (RPS) (Safwat Kabel and Bassim (2019). As part of the government's policy, Boie (2016) distinguished production and investment incentives.

Technological changes are the key to the energy transformation of countries. de Coninck and Puig (2015) demonstrated the importance of research conducted by institutions and companies on technical development of renewable energy. Xia et al. (2020), on the example of the photovoltaic industry, showed the importance of state support for research and development projects. New technical solutions allow to reduce production costs, and thus increase their competitiveness. According to Chu and Majumdar (2012), technology costs are the key to use of RES on a larger scale.

de Coninck and Puig (2015) emphasized the importance of technological diffusion in the development of renewable energy sources, especially in developing countries. An important factor in this process is social acceptance (Wüstenhagen et al. 2007;

Nkundabanyanga et al. 2020). Research shows that people's acceptance of RES technology is influenced by the features of the technology (e.g., reliability, benefits, savings), the business environment, and psychological, social and institutional factors (Islam 2014; Kardooni et al. 2016).

Suzuki (2015) emphasized the role of international institutions in the field of cooperation within research and development programs, financial support and creating an environment conducive to technological diffusion in the field of renewable energy.

3 Directions and Mechanisms of Financing the Green Energy Transformation

The transition to a low-carbon and ultimately zero-carbon economy is a challenge facing modern countries. The pace and way of achieving this goal depends on the adopted state energy strategy and on the individual behavior of economic entities and households. Differences in environmental conditions and natural resources owned by individual countries, in technological advancement and infrastructure, as well as the type and scope of state support in the implementation of green energy projects mean that the structure of generated energy and the share of renewable energy in the total energy of individual countries are different (Figs. 1, and 2).

The energy transformation is a complicated process that requires large investment capital. Investments in this area may be financed from public and/or private funds. The approach of public and private sector entities to implemented projects as well as the sources and possibilities of obtaining financing are different. The private sector tends to focus on return-on-investment projects, while public sector investment is generally designed to support the development of renewable energy in specific regions by, for example, reducing the cost of capital. The main source of financing renewable energy projects is the private sector. This sector financed over 80% of all investments carried out in years 2013–2018 (Fig. 3).

Private investment entities include project developers, non-energy-producing companies (corporate actors), commercial financial institutions, households, institutional investors and private equity, venture capital and infrastructure funds (Fig. 4).

In 2013–2018, project developers were the main entities in the private sector implementing investment in renewable energy. Their share in investment in years 2013–2018 was systematically increasing and amounted 55% in 2018. Commercial financial institutions (i.e., commercial and investment banks) are the second largest group of private sector entities involved in the implementation of renewable energy projects. Their contribution has been estimated at over 20% of total private funding in years 2014–2018.

In recent years, non-energy-producing companies have played an important role in the energy transformation, allocating in RES investments an average of USD 17 billion annually in years 2017–2018. Contrary to project developers who distribute

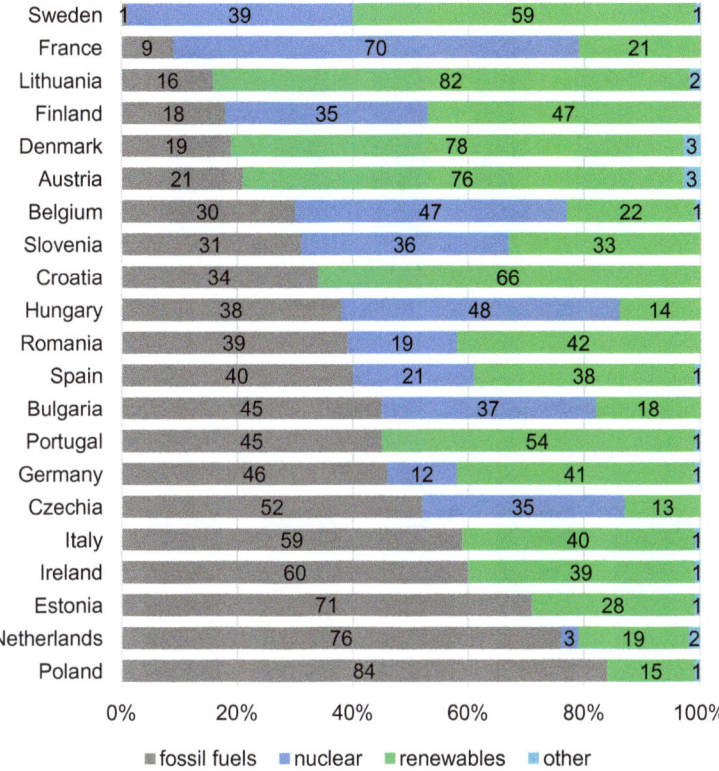

Fig. 1 Sources of Energy in European countries, 2020, %. *Source* American Chamber of Commerce in Poland (2020) Energy Transformation in Poland. AmCham Business and Economics Review, 1/2022

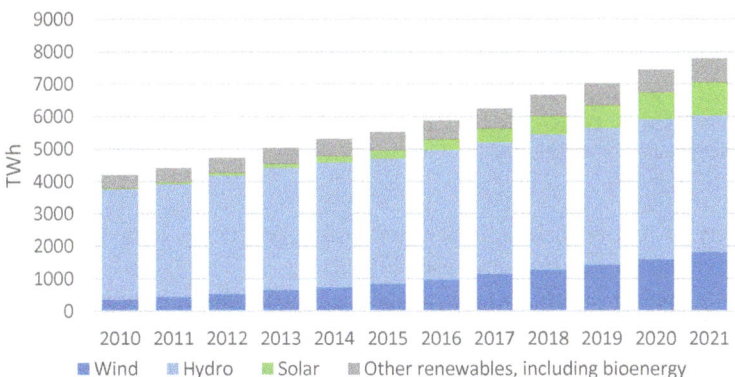

Fig. 2 Electricity generated from renewable sources, 2010–2021, TWh. *Source* Own elaboration based on Our World in Data database. https://ourworldindata.org/grapher/modern-renewable-prod

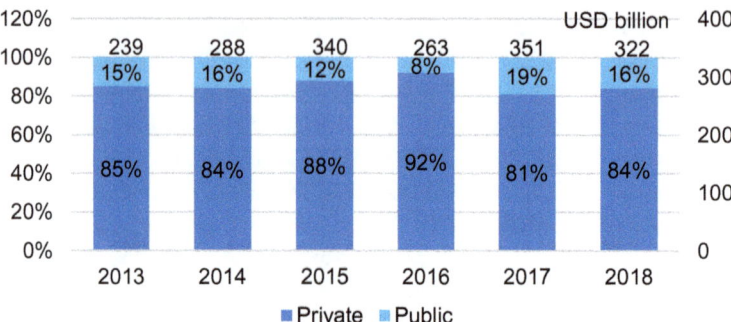

Fig. 3 Private and public investment in renewable energy finance, 2013–2018. *Source* Own elaboration based on: IRENA and CPI (2020), Global Landscape of Renewable Energy Finance, 2020, International Renewable Energy Agency, Abu Dhabi

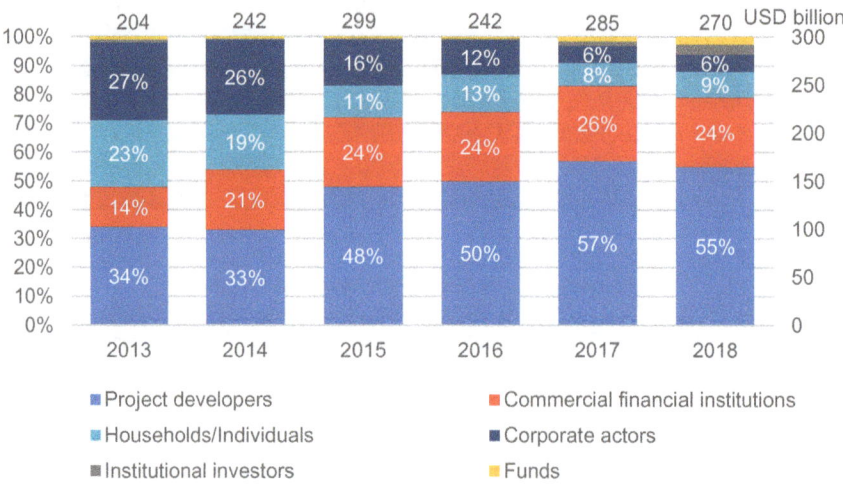

Fig. 4 Private investment in renewable energy, by investment source, 2013–2018. *Source* Own elaboration based on: IRENA and CPI (2020), Global Landscape of Renewable Energy Finance, 2020, International Renewable Energy Agency, Abu Dhabi

electricity as part of their activities, non-energy-producing companies most often generate electricity for their own use (CPI 2012).

Institutional investors, which include pension funds, insurance companies, sovereign wealth funds, endowments and foundations manages, are a group of entities that have the potential to play a key role in channeling capital into new renewable energy projects (directly or indirectly), and in refinancing existing assets of renewable energy (e.g., through green bonds) to free up capital for new investments. Research by IRENA shows that in the last two decades around 20% of investment have been realized indirectly through the funds focused on renewable energy sources, while only

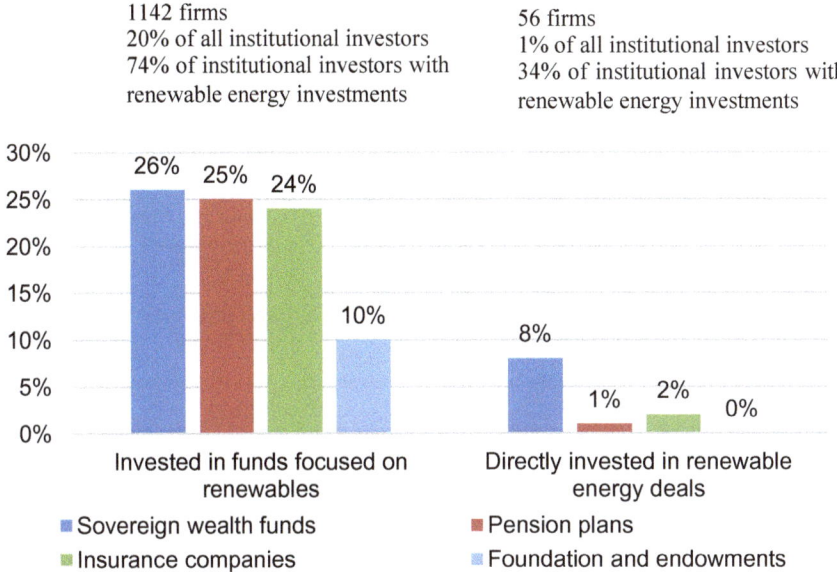

Fig. 5 Number of institutional investors with investments in renewable energy

around 2% of such institutions have invested directly in renewable energy projects (Fig. 5).

More than 75% of all direct investments made by institutional investors in renewable energy projects between 2009 and 2019 were transactions at the secondary stage, i.e., investments in existing assets that do not require further financing. Institutional investors prefer to invest in wind power because this technology has proven successful.

Investments in renewable energy can also be made from public sources. In this way, a small percentage of investments is financed (an annual average of around 14% of total investments in years 2013–2018). Governments and state agencies, investments from DFIs, and national and multilateral climate funds, constitute public finance for RE sector. Contrary to the private sector, where the basic criterion for the decision to implement a RES project is the investment profitability criterion, in the case of the public sector, the goal is to implement investments in sectors and regions that are relatively immature or difficult to invest. These could be, for example, projects to increase access to clean energy for households off-grid, or to implement investments in urban infrastructure (e.g., public lighting).

RES-related projects are infrastructure projects and as such can be implemented as project finance or corporate finance. Project finance is based on the assumption that the basis for obtaining financing is not the creditworthiness of sponsors, but the ability of the project itself to generate cash flow allowing to repay debt and compensate capital. In this case, the provider of external capital will be, depending

Table 1 Alternative sources of financing renewable energy projects

Category	Instrument	Project finance	Corporate finance
Debt	Bonds	Project Bonds Green Bonds	Corporate Bonds Green Bonds
	Loans	Syndicated loans Direct Lending (to project)	Direct lending (to corporate) Sybdicated and securitized loans
	Hybrid	Subordinated Debt Mezzanine Finance	Subordinated Bonds Convertible bonds
Equity	Listed	YieldCos	Listed Stocks, etc.
	Unlisted	Direct investment project (SPV) Equity	Direct Investment in Corporate (SPV) Equity

Source Vazquez (2018). Financing the transition to renewable energy in the European Union, Latin America and the Caribbean

on the financial system (bank-based and market-based financial system), banks or capital market entities, such as e.g., an investment fund.

Corporate finance is a traditional method of financing infrastructure projects, especially those carried out by the private sector. Firms in charge of the infrastructure (i.e., building and operating projects) issue shares or borrow on capital markets in order to obtain financing. Such companies often have a portfolio of projects. On energy markets, utilities typically have a portfolio of energy projects with different risk profiles. Table 1 presents a comparison of financing sources in the case of project finance and corporate finance projects.

The main instruments for financing infrastructure projects are loans and bonds. Debt markets enable the creation of maturity products that are consistent with the lifetime of the infrastructure project, that is long-term. Debt instruments may also benefit from investors who prefer long-term investments. An example of such companies are pension or insurance funds.

An important part of debt instruments is subordinated debt, the purpose of which is to absorb credit losses before senior debt. Thus, it increases the quality of the senior debt. the advantage here is the ability to design subordinated debt to carry different risk/ rates of return that bridge between traditional debt and equity.

Equity financing can also be seen as project venture capital (usually required to start a project or refinance it). Listed shares would be traded on public markets, while unlisted shares would provide direct control of the project. Project equity financing may be located closer to debt instruments in the sense that infrastructure contracts may generate relatively low risk/return ratios.

One of the examples of reducing/separating the risk related to the implementation of renewable energy projects are YieldCos. They are an asset class of publicly traded companies that are focused on returning cash flows generated from renewable energy assets to shareholders. In the energy industry, public utility companies implement renewable energy projects in the operational phase through subsidiaries and issue shares on public (listed) markets. An example of such a company is SunEdison.

Financing instruments in RES cover the technological development phase and the project implementation phase. The first group of instruments includes programs created and implemented by international institutions. An example is the financing programs for renewable energy projects created by the European Commission. Most of these programs focus on projects in the research and development or pilot plant phase. Examples of such programs are: Horizon 2020, The European Research Framework Program, The Marguerite Fund, Global Energy Efficiency and Renewable Energy Fund (GEEREF), European Local Energy Assistance (ELENA), New Entrant Reserve 300 (NER 300). Apart from public financing, this support covers projects implemented in the public–private partnership formula and projects financed with private funds.

Traditionally, the implementation of renewable energy projects was based on bank sources. Due to the confidentiality of concluded banking agreements, it is difficult to obtain statistical data in this regard. However, the analyzes of IRENA and CPI (2020) show that RES projects are financed mainly with loans. In the case of projects implemented by private sector entities, these are usually commercial banks.

Green energy projects can get support from various international financial institutions. These include the European Investment Bank (EIB), the European Bank for Reconstruction and Development (EBRD) and the UK Green Investment Bank (GIB). Support instruments are different. EIB for projects with usually accepted risk offers medium and long-term loans with fixed or variable interest rates. Depending on the size of the project, it either provides individual loans or indirect loans through intermediary partner banks. For projects with higher-than-standard risk, a support is provided through Structured Finance Facility (SFF), such as Senior loans and guarantees incorporating pre-completion and early operational risk; Subordinated loans and guarantees ranking ahead of shareholder subordinated debt; Mezzanine finance, including high-yield debt for small and medium enterprises (SMEs) experiencing high-growth or which are undergoing restructuring and project-related derivatives (Vazquez 2018). In turn, Green Investment Bank (GIB) provides flexible capital, covering investment across the full capital structure, including debt, mezzanine debt, and equity. Support for major projects is provided directly, while smaller projects are indirectly supported through funds or developer partnerships (Vazquez 2018). Investors in the renewable energy market obtain financing from capital market instruments, such as green bonds, more and more often. The green bond market has been developing dynamically for several years. In 2019, its value increased by more than a half, i.e., from USD 179 trillion to USD 271 trillion. Over 50 per cent of green bonds are bonds related to renewable energy sources (Fig. 6).

According to IRENA, between 2007 and 2019, the average value of the issue of renewables-dedicated green bonds was USD 364 million, which means that they are higher than the issue of other environmental bonds. For this reason, renewables-dedicated green bonds become particularly attractive to larger investors (e.g., institutional investors) who reduce their unit costs by entering into larger transactions. Through investing in renewables-dedicated green bonds, investors can lower their investment risk by gaining access to a diversified portfolio of already operating renewable energy assets.

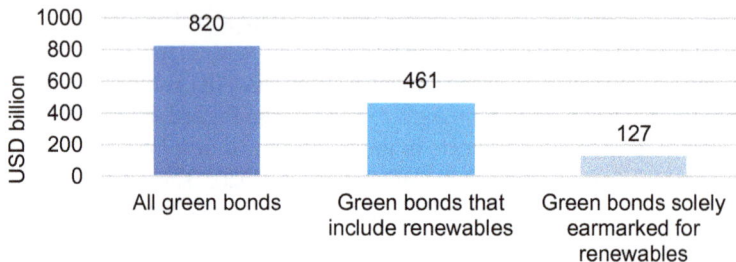

Fig. 6 Volume of all green bonds, green bonds that include renewable energy and renewables-dedicated green bonds, 2007–2019. *Source* IRENA and CPI (2020), Global Landscape of Renewable Energy Finance, 2020, International Renewable Energy Agency, Abu Dhabi, p. 38

Due to the capital-intensive nature of the energy transformation process, a very important element influencing the speed and quality of the ongoing changes is the type and scope of state support for the implementation of renewable energy projects. Many countries are developing a renewable energy policy for this purpose (Rennkamp et al. 2017). Donastorg et al. (2017) showed that this policy plays a key role in promoting innovation in area of renewable energy technologies and contributes to costs reduction, thereby increasing the competitiveness of renewable energy compared to traditional energy. Renewable energy policies include financial incentives and renewable portfolio standards (Table 2).

The types of instruments of renewable energy policies applied by individual countries depends on the implemented energy policy strategy. An overview of instruments of renewable energy policies in selected countries is presented in Table 3.

Xin-gang et al. (2017) showed that the renewable energy policies that are most used by countries are Feed-In Tariff (FIT) and Renewable Portfolio Standard (RPS). Yang et al. (2021) noted that FIT is suitable for the early stages of development of the renewable energy industry. As the industry grows, FIT and RPS can be combined to ensure healthy and sustainable development. In turn, Choi et al. (2018) compared the performance of FIT and RPS from the perspective of the government and renewable energy producers. Research shows that RPS was more efficient for solar PV from a government perspective, while FIT was more efficient for non-photovoltaic energy, such as wind power, bioenergy, and fuel cells.

4 Case Studies of the Implementation of Financing the Green Energy Transformation. Transition to Green Energy in Sweden

Main energy sources used in Sweden are: nuclear fuel, biomass and oil (crude and oil products). These sources together provided more than 75% of total energy supplied

Table 2 Instruments of renewable energy policy

Instruments of renewable energy policy			
Financial instrument	Tax incentives		By reducing the tax burden, states encourage investment in renewable energy. Examples of tax incentives include: tax deductions, tax exemption, and tax credit, tax refunds, rebates on taxes, and the tax exemptions
	Loan		Through loans granted on preferential terms or government loan programs, governments reduce the cost of raising capital and thus encourage the implementation of renewable energy projects. Most often these are loans with low or zero-interest rates, and loans programs such as low-interest rates, longer amortization, low hassle and administrative fees, unsecured loans aimed to increase investment in renewable energy
	Feed-in-tariff		One of the most widely used financial instruments supporting the development of renewable energy sources. Haselip (2011) showed that the feed-in tariff, reducing the financial risk associated with individual RES projects, encourages investors. Nicolini and Tavoni (2017) calculated that the feed-in-tariff increase by 1% contributes to an increase in motivated renewable production by 0.4–1%
Renewable Portfolio Standard (RPS)			It is a scheme that requires energy suppliers to supply a certain amount of their renewable electricity portfolio (LCA 2006). By increasing the price that a producer of renewable energy will receive to produce it, the Renewable Portfolio Standard (RPS) helps to encourage the production of renewable energy

Source Own elaboration base on Safwat Kabel and Bassim (2019)

in 2019 (Fig. 7). Also hydropower has a significant meaning for energy supply in Sweden.

Over last 40 years the supply of biofuels has tripled while the supply of crude oil and petroleum products decreased more than half (Fig. 8). The share of nuclear fuel remains at a similar level demonstrating some periodic fluctuations. Also, the share of hydropower is quite stable and ranks second among renewable energy sources. The lowest share has solar power, however, in the last few years more and more photovoltaic (PV) panels are being installed. In years 2019–2020 the number of PV systems connected to power grid increased by 50%.

Electricity in Sweden is mainly generated by hydropower and nuclear power, however, there has been a continuous increase in wind power, which is currently the third largest source of electricity generation.

The share of renewable energy in Sweden has been at a high level for many years. In 1990, a third of the energy came from renewable sources, and in 2019, renewables provided over 56% of the energy (Fig. 9).

Despite the relatively high share of renewable energy in Sweden, the country plans to continue increasing it. In fact, such high share of renewables comes mainly from

Table 3 Instruments of renewable energy policy

Country	Financial incentives			R&D incentives	Market incentives	Incentives via regulations		
	Tax Support	Financial Subsidy/Loan	Feed-In-Tariffs (FIT)	R&D support	Market regulation	Green certificates	Power purchase legislation	Defined targets
Australia	✓	✓	✓			✓	✓	✓
Belgium			✓					
Brazil	✓	✓	✓			✓	✓	
China	✓	✓	✓	✓	✓			
Denmark		✓	✓					
France		✓						
Germany		✓	✓		✓			
India	✓	✓	✓			✓	✓	
Italy		✓	✓					
Montenegro		✓						
Morocco		✓					✓	
Sweden		✓				✓		
UK		✓				✓		
USA	✓	✓	✓			✓	✓	

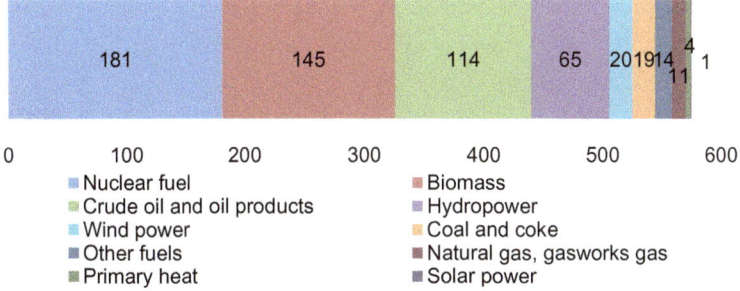

Fig. 7 Energy supply in Sweden in 2019, TWh. *Source* Own elaboration based on: Energy in Sweden 2021, An overview. Swedish Energy Agency. https://energimyndigheten.a-w2m.se/Home.mvc?ResourceId=198022

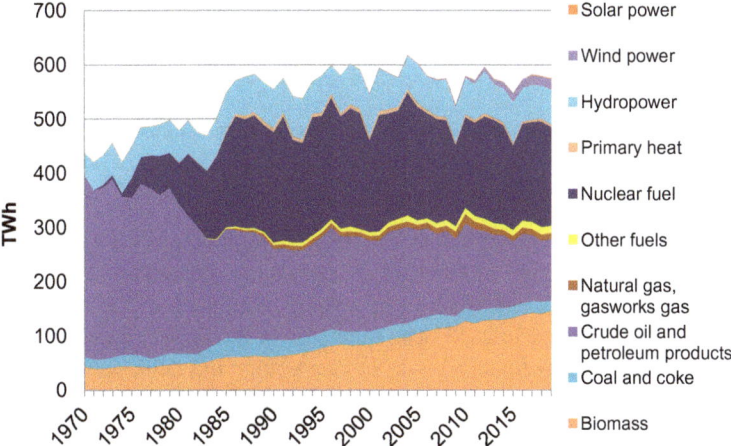

Fig. 8 Total energy supply by energy commodity, TWh. *Source* Own elaboration based on: Energy in Sweden 2021, An overview. Swedish Energy Agency. https://energimyndigheten.a-w2m.se/Home.mvc?ResourceId=198022

biomass and hydropower. When it comes to wind and solar power there is still a lot to improve.

Sweden's Integrated National Energy and Climate Plan describes several objectives related to emission reduction and energy generation and use. One of the targets is 100% renewable electricity generation by 2040. This target seems to be ambitious but can be achieved by increasing the share of wind and solar power from today's 11–42% (39% wind and 3% solar) in 2040. To achieve this target, a combination of several elements is needed. The most important of them are:

1. construction of new windfarms and solar PV systems, along with necessary infrastructure;

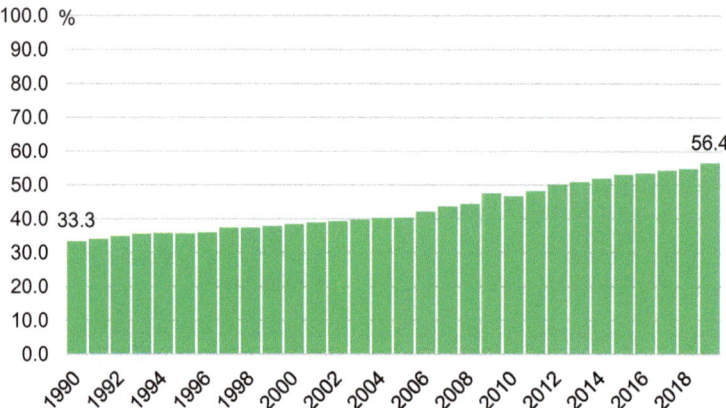

Fig. 9 Share of renewables in energy use in Sweden, %. *Source* Own elaboration based on: Energy in Sweden 2021, An overview. Swedish Energy Agency. https://energimyndigheten.a-w2m.se/Home.mvc?ResourceId=198022

2. Innovative technologies allowing to increase energy efficiency and to decrease energy use;
3. Changes in energy market design.

Among other challenges in achieving 100% renewable electricity generation is financing of investment and research in area of energy generation. An example of a large investment is the construction of a wind farm, such as the 203 MW wind farm in Jädraås, launched in 2013. The total cost of the project was approximately EUR 360 million. It was financed with private equity and debt, part of which was government-backed (Fig. 10). The equity in the amount of EUR 120 million, came from Swedish largest windfarm developer, Arise Windpower, and UK private investor Platina Partners. Another EUR 120 million was provided as a debt by two commercial banks (DNB and SAB), and the remaining EUR 120 million by Danish pension fund PensionDenmark. The latter one was guaranteed by Danish export credit agency, which was backed by the Kingdom of Denmark. The case of Jädraås wind farm shows that state-backed financing is crucial for large-scale renewable energy projects (Lam and Law 2018). Thanks to the Swedish policy of transition to renewable energy, investors can count on long-term co-financing of projects by the government and government-backed debts. Along with the availability of expert knowledge and effective management of relations between stakeholders, it allows to reduce the risk borne by investors.

The important role of government support is also confirmed by another project, the 372 MW Björnberget wind farm. The farm is expected to be in operation by the end of 2022. Financing of the project was provided by a consortium of lenders including banks and state-owned Swedish Export Credit Corporation (SEK).

Another objective of Swedish energy and climate policy is to cut net greenhouse gas emissions to zero and then achieve negative emissions by 2045. One of the main challenges in achieving this goal is decarbonisation of Swedish heavy industry which

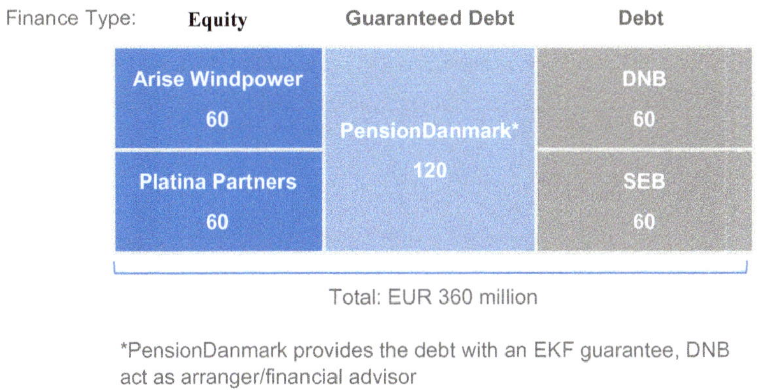

Finance Type:	**Equity**	**Guaranteed Debt**	**Debt**
	Arise Windpower 60	PensionDanmark* 120	DNB 60
	Platina Partners 60		SEB 60

Total: EUR 360 million

*PensionDanmark provides the debt with an EKF guarantee, DNB act as arranger/financial advisor

Fig. 10 Structure of financing Jädraås Onshore Windfarm (amount in EUR millions). *Source* Own elaboration based on: Boyd and Hervé-Mignucci (2013)

is responsible for almost one-third of Sweden's greenhouse gas emission (about 17 Mt of CO_2 per year) (Nykvist et al. 2020). It is estimated that investment increase of additional about EUR 6.5 billion is needed to decarbonize the four largest emission sources in Sweden: steel production, cement production, petrochemical industry, and oil refining. These sources represent about 70% emission from industry (12 Mt of CO_2 per year). The estimated EUR 6.5 billion does not include investment needed to maintain current production levels.

The low-carbon transition in heavy industry will require substantial and long-term support mechanisms. Companies expect significant risk-sharing and investment support from the government, but it is unlikely that it will take on most of the financial burden of transition (Nykvist et al. 2020).

Transition to low-carbon production requires capital-intensive investment but also increases production cost due to use of different technologies. In the assessment of the cost of production there are the two main categories of costs:

1. Capital expenses (CAPEX)—the investments needed to build new production facilities, upgrade existing facilities, and/or purchase new equipment,
2. Operating expenses (OPEX)—the difference between the operating costs of existing production processes and new processes, using new or upgraded technology.

The shares of CAPEX and OPEX in product cost for four discussed industries are presented in Table 4. Presented values are the results of analysis and calculations made by researchers from Stockholm Environment Institute. The results show that CAPEX a more significant challenge in steel and petrochemicals than in cement and refining. The OPEX challenge dominates in the refining because most of the same processes remain in the low carbon case, which means that CAPEX costs for the transition are limited. Additional OPEX cost in cement industry, is much higher than CAPEX costs.

Table 4 Summary of the importance of the CAPEX challenge across four industries

Economic barrier driven by CAPEX challenge	Petrochemicals (%)	Steel (%)	Refining (%)	Cement (%)
Share of abatement cost that is CAPEX	18	40	20	7
Product cost increase due to CAPEX	7	4	1	6
Product cost increase due to OPEX	31	7	3	86

The Swedish government provide a Industriklivet program aimed at supporting investments in decarbonizing industry. It offers 600 million SEK per year, however, based on their study, experts from Stockholm Environment Institute suggest doubling or even tripling direct public support throughout the next decade.

5 Summary

Energy transformation is one of the most important challenges of modern times. The transition to a low-emission and ultimately zero-emission economy requires cooperation between market players at various levels: local, national and international, hence we can distinguish initiatives taken at the EU level, renewable energy policy of individual countries, and activities carried out by specific enterprises or households.

National policies are crucial for the effective transition from an economy based on fossil fuels to technologies using renewable energy sources. It requires the adoption and implementation of measures that take into account a wide range of stakeholders. On the one hand, through the renewable energy policy instruments, entities should be stimulated to invest in renewable energy, and on the other hand, support for conventional energy producers should be provided. Due to the capital intensity of the energy transformation process, the adopted model of financing investments in renewable energy projects is of particular importance, i.e., the definition of financing sources for renewable energy projects implemented by the private and public sectors and the creation of a regulatory framework for projects implemented in the public–private partnership formula. It is worth noting that public funds should be subsidiary to private market sources of financing.

The vast majority of renewable energy projects require obtaining external capital. Depending on the financial system, its source can be banks or the capital market, that has been developing very dynamically in recent years in the green bonds segment. The use of green bonds also helps to reduce the risks associated with investments in RES.

The risk accompanying the implementation of renewable energy projects is relatively high, which also affects the cost of capital, which is why renewable energy policy instruments such as tax incentives, loan, feed-in-tariffs and Renewable Portfolio Standard (RPS) are extremely important. Their use depends on the type of renewable energy the project relates to, the size of the project and the entity implementing it.

There is no single proper strategy for financing the country's energy transformation. This is confirmed by examples of countries that pursue the goal of a low carbon economy in different ways. China based its energy transformation on a state-controlled policy, while German policy engages all stakeholders, including households and small energy companies, to invest in renewable energy. Meanwhile, in the UK, the energy transition to RE is based on energy market players.

References

Aldieri L, Vinci CP (2018) Green economy and sustainable development: the economic impact of innovation on employment. Sustainability 10(10):3541

Bina O, La Camera F (2011) Promise and shortcomings of a green turn in recent policy responses to the "double crisis". Ecol Econ 70(12):2308–2316. S0921800911002680. https://doi.org/10.1016/j.ecolecon.2011.06.021

Boie I (2016) Determinants for the market diffusion of renewable energy technologies—an analysis of the framework conditions for non-residential photovoltaic and onshore wind energy deployment in Germany, Spain and the UK. Dissertation for the Doctoral Degree. University of Exeter, Devon

Boyd R, Hervé-Mignucci M (2013) San Giorgio Group case study: Jädraås Onshore Windfarm, CPI Report

Brown E et al (2014) Green growth or ecological commodification: debating the green economy in the global South. Geogr Ann: Ser B Hum Geogr 96(3):245–259

Choi G, Huh SY, Heo E, Lee CY (2018) Prices versus quantities: comparing economic efficiency of feed-in tariff and renewable portfolio standard in promoting renewable electricity generation. Energy Policy 113:239–248

Chu S, Majumdar A (2012) Opportunities and challenges for a sustainable energy future. Nature 488(7411):294–303. https://doi.org/10.1038/nature11475

Conti C, Mancusi ML, Sanna-Randaccio F, Sestini R, Verdolini E (2018) Transition towards a green economy in Europe: innovation and knowledge integration in the renewable energy sector. Res Policy 47(10):1996–2009

CPI (2012) Global landscape of climate finance 2012, climate policy initiative, London. https://www.climatepolicyinitiative.org/wpcontent/uploads/2012/12/The-Landscape-of-ClimateFinance-2012.pdf

de Coninck H, Puig D (2015) Assessing climate change mitigation technology interventions by international institutions. Clim Change 131(3):417–433

Donastorg A, Renukappa S, Suresh S (2017) Financing renewable energy projects in developing countries: a critical review. IOP Conf Ser: Earth Environ Sci 83(1)

Energy in Sweden (2021) An overview. Swedish Energy Agency. https://energimyndigheten.a-w2m.se/Home.mvc?ResourceId=198022

Haselip JA (2011) FIT for use everywhere? Assessing experiences with renewable energy feed-in tariffs. In diffusion of renewable energy technologies: case studies of enabling frameworks in developing countries. UNEP Risø centre on energy, climate and sustainable development. Department of management engineering. Technical University of Denmark (DTU). Technology Transfer Perspectives Series, pp 89–100

Hickel J, Kallis G (2020) Is green growth possible? New Polit Econ 25(4):469–486

IRENA and CPI (2020) Global landscape of renewable energy finance. International Renewable Energy Agency, Abu Dhabi

Islam T (2014) Household level innovation diffusion model of photovoltaic (PV) solar cells from stated preference data. Energy Policy 65:340–350

Jacobs M (1991) The green economy: environment, sustainable development, and the politics of the future. University of California: Pluto Press

Kardooni R, Yusoff SB, Kari FB (2016) Renewable energy technology acceptance in Peninsular Malaysia. Energy Policy 88:1–10

Kasayanond A (2019) Environmental sustainability and its growth in Malaysia by elaborating the green economy and environmental efficiency. 670216917

Koval V, Sribna Y, Kaczmarzewski S, Shapovalova A, Stupnytskyi V (2021) Regulatory policy of renewable energy sources in the European national economies. Polityka Energetyczna 24

Lam PTI, Law AOK (2018) Financing for renewable energy projects: a decision guide by developmental stages with case studies. Renew Sust Energ Rev 90:937–944

Lavrinenko O, Ignatjeva S, Ohotina A, Rybalkin O, Lazdans D (2019) The role of green economy in sustainable development (case study: the EU states). Entrepreneurship Sustain Issues 6(3):1113

LCA (2006) Analysis of a renewable portfolio standard for the state of North Carolina. La Capra Associates, Boston

Liu Y, Dong F (2021) How technological innovation impacts urban green economy efficiency in emerging economies: a case study of 278 Chinese cities. Resour Conserv Recycl 169:105534

Merino-Saum A, Clement J, Wyss R, Baldi MG (2019) Unpacking the green economy concept: a quantitative analysis of 140 definitions. J Cleaner Prod 118339. https://doi.org/10.1016/j.jclepro.2019.11833

Mikhno I, Koval V, Shvets G, Garmatiuk O, Tamošiūnienė R (2021) Green economy in sustainable development and improvement of resource efficiency. Cent Eur Bus Rev (CEBR) 10(1):99–113

Nicolini M, Tavoni M (2017) Are renewable energy subsidies effective? Evidence from Europe. Renew Sustain Energy Rev 74:412–423. S1364032116310905. https://doi.org/10.1016/j.rser.2016.12.032

Nkundabanyanga SK, Muhwezi M, Musimenta D, Nuwasiima S, Najjemba GM (2020) Exploring the link between vulnerability of energy systems and social acceptance of renewable energy in two selected districts of Uganda. Int J Energy Sect Manag

Nykvist B, Maltais A, Olsson O (2020) Financing the decarbonisation of heavy industry sectors in Sweden. Stockholm Sustainable Finance Centre

Pearce DW, Anil M, Barbier E (1989) Blueprint for a green economy, vol 1. Earthscan

Regulation (EU) 2020/852 of the European Parliament and of the Council of 18 June 2020 on the establishment of a framework to facilitate sustainable investment and amending Regulation (EU) 2019/2088 (Text with EEA relevance) renewable portfolio standard_A case study of China's waste incineration power industry. Waste Manag 68:711–723

Rennkamp B, Haunss S, Wongsa K, Ortega A, Casamadrid E (2017) Competing coalitions: the politics of renewable energy and fossil fuels in Mexico, South Africa and Thailand. Energy Res Soc Sci 34:214–223

Safwat Kabel T, Bassim M (2019) Literature review of renewable energy policies and impacts. KABEL, Tarek Safwat, pp 28–41

Stoknes PE, Rockström J (2018) Redefining green growth within planetary boundaries. Energy Res Soc Sci 44:41–49

Suzuki M (2015) Identifying roles of international institutions in clean energy technology innovation and diffusion in the developing countries: matching barriers with roles of the institutions. J Clean Prod 98:229–240

The Global Risks Report (2021) https://www.weforum.org/reports/the-global-risks-report-2021. Access: 6.02.2022

Vazquez M (2018) Financing the transition to renewable energy in the European Union, Latin America and the Caribbean

Wüstenhagen R, Wolsink M, Bürer MJ (2007) Social acceptance of renewable energy innovation: an introduction to the concept. Energy Policy 35(5):2683–2691

Xia S, Ding Z, Du T, Zhang D, Shahidehpour M, Ding T (2020) Multitime scale coordinated scheduling for the combined system of wind power photovoltaic thermal generator hydro pumped storage and batteries. IEEE Trans Ind Appl 56(3):2227–2237. 9000510. https://doi.org/10.1109/TIA.2020.2974426

Xin-Gang Z, Yu-Zhuo Z, Ling-Zhi R, Yi Z, Zhi-Gong W (2017) The policy effects of feed-in tariff and renewable portfolio standard: A case study of China's waste incineration power industry. Waste Manage 68:711–723

Yang DX, Jing, YQ, Wang C, Nie PY, Sun P (2021) Analysis of renewable energy subsidy in China under uncertainty: feed-in tariff versus renewable portfolio standard. Energy Strategy Rev 34:100628

Ying L, Li M, Yang J (2021) Agglomeration and driving factors of regional innovation space based on intelligent manufacturing and green economy. Environ Technol Innov 22:101398

Yumei H, Iqbal W, Irfan M, Fatima A (2022) The dynamics of public spending on sustainable green economy: role of technological innovation and industrial structure effects. Environ Sci Pollut Res 29(16):22970–22988

Zhang D, Mohsin M, Rasheed AK, Chang Y, Taghizadeh-Hesary F (2021) Public spending and green economic growth in BRI region: mediating role of green finance. Energy Policy 153:112256

Zhao S, Jiang Y, Wang S (2019) Innovation stages, knowledge spillover, and green economy development: moderating role of absorptive capacity and environmental regulation. Environ Sci Pollut Res 26(24):25312–25325

Zhou YH, Pu YL, Chen SY, Fang F (2015) Government support and development of emerging industries: a new energy industry survey. Econ Res J 50(6):147–160

Green Energy Transformation Models—Main Areas and Further Directions of Development

Katarzyna Cheba and Iwona Bąk

Abstract Increasing the share in the so-called energy mix of pure energy is a part of the EU investment plan regarding the European Green Deal. There is the growing importance of innovations in the field of renewable, clean and green technologies, and the prospect of a green economy raises the requirements of the market in respect of ecological solutions. Experts stress that changes occurring in the energy sector should be based on the 3D model, i.e. decarbonisation, decentralisation and digitalisation. The pace of the green transformation is also important, and is highly diversified depending on the level of the economic development of individual countries. The main purpose of the research presented in this chapter was to identify the main areas comprising a complex model (models) of green transformation in EU countries, and to indicate potential directions of changes in this transformation. In the paper a TOPSIS method was used for this purpose. The analyses suggest that it is practically impossible to construct one or even a few models of green transformation for countries in the EU (and United Kingdom). The results of individual countries were strongly diversified.

1 Introduction

Green transformation has now become one of the more important subjects of research addressed in publications by authors representing diverse disciplines of science (i.e. Borel-Saladin and Turok 2013; Amundsen and Hermansen 2021).

Studies concerning this area cover theoretical issues linked with defining this term (Crespi et al. 2016; Cui and Lui 2021), and research papers presenting, among others, analyses of the current level of green transformation as well as potential directions of change in this scope (Feng and Chen 2018; Declich et al. 2020). Both approaches

K. Cheba (✉) · I. Bąk
Faculty of Economics, West Pomeranian University of Technology in Szczecin, Szczecin, Poland
e-mail: katarzyna.cheba@zut.edu.pl

I. Bąk
e-mail: iwona.bak@zut.edu.pl

are crucial for understanding the essence of green transformation and preparing the most important assumptions for the models focused, above all, on green energy transformation.

The aim of the considerations presented in this chapter was to identify the main areas comprising a complex model (models) of green transformation in EU countries, and to indicate potential directions of changes in this transformation.

The starting point for building models in this scope was a systematic literature review, based on papers indexed in the Web of Science (WoS) and Scopus databases. The main direction in research and analyses of the assembled publications was to find areas (groups of factors, conditions) considered by various authors (see Feola 2015; Gea-Bermúdez et al. 2021) when defining the term 'green transformation'.

Although the main area of considerations presented in this chapter regards green energy transformation, understood as the transition from the use of non-renewable sources of energy (mostly coal, crude oil, and natural gas) towards renewable sources (wind energy, as well as solar, water and geothermal energy coming from inside the Earth), the authors also noted the evolution of the approach to that term. This evolution is visible in a gradual transition from the market model, in which the main determinants deciding about the attitude of society (businesses, citizens, and the authorities) to the applied rules of the use of energy are: demand, supply, price, effectiveness and competitiveness (Devine-Wright and Murphy 2007; Schot et al. 2016), towards the civic model which places importance on the impact of consumer behaviour and purchasing decisions made by citizens on the transition towards the economy which is environmentally and socially responsible, sustainable, with low emissions, 'green' and supported by 'clean' technologies (Lennon et al. 2019).

It is worth noting here that even though the term 'green transformation' is widely used, above all in the public debate, yet the subject literature lacks comprehensive analyses in this area. The existing studies are mainly of a theoretical nature, and concentrate on defining the term, some merely present research results regarding specific technologies or analyses of singular conditions of this process. This chapter attempted to integrate these two approaches. The first part, while reviewing the literature, also describes different ways of defining green transformation and indicates key conditions of this process. The second part, using an empirical example, compares the achievements of EU countries (and also United Kingdom) in terms of green transformation. The information obtained from the conducted review of the subject literature was also used to point towards the main areas and potential directions of changing the model/s of green transformation in the EU.

2 Green Transformation, Its Main Areas and Potential Directions of Change—Theoretical Perspective

Green transformation is defined in the subject literature in a variety of ways. The Web of Science and Scopus databases now comprise over 750 publications whose titles, keywords and/or abstracts contain references to the terms 'green transformation' and or 'green transition'. More than 270 of them also describe models of this kind of transition, while 51 make direct reference to the terms 'green energy transformation' and 'green energy transition' (see Table 1). A growing interest in research on these issues can be noted especially during the last five years. In 2021 there were 104 publications, and a joint number of citations of all those related to the subject amounted to 1232. From the beginning of 2022 (the research was carried out at the end of April 2022), the WoS and Scopus databases included 43 studies, while all the publications referring to issues of green economy were quoted 556 times in 2022 alone (Fig. 1).

The analysis of the published works suggests that green transformation is defined in the literature in a variety of ways. Some authors (Bjørner and Jensen 2002), describe this kind of transformation mostly from the perspective of policy, i.e. actions undertaken by the authorities (the institutional perspective). In publications by other

Table 1 Number of papers identified in the WoS and Scopus databases according to the selected keywords

The combinations of topic	Number of papers
"Green transformation*"	461
"Green transition*"	304
"Green transformation*" OR "green transition*"	751
"Green transformation*" AND model*	176
"Green transition*" AND model*	105
"Green transformation*" AND model* OR	
"Green transition*" AND model*	272
"Green energy* transformation*"	4
"Green energy* transition*"	47
"Green energy* transformation*" OR "Green energy* transition*"	51

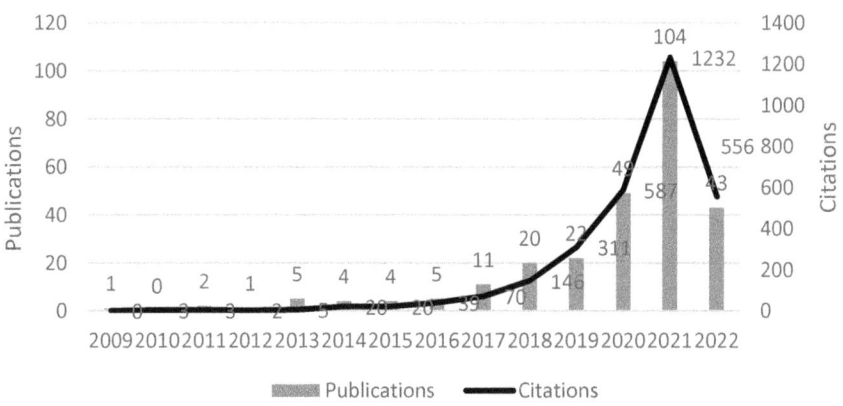

Fig. 1 Total publications and citations by year

authors, e.g. Rosenow et al. (2017), Chou and Mathews (2017), Wang et al. (2018), and Hu et al. (2021), the main subject is the role of the government in the process of energy transformation. These studies attempted to evaluate the effectiveness of policies from the viewpoint of their impact on the production and use of energy. Some (i.e. Ruszel 2017; Rosenow et al. 2017), concentrate on comparative analyses of policies implemented by different countries. They also discuss the influence of the ways of collecting environmental taxes and their rates on the process of green transformation (Ziolo et al. 2020). Practices applied by European countries differ, while in the majority of them the level of the environmental taxes depends on, among others, the levels of pollution emitted into the environment (in line with the principle that "who pollutes, pays"). However, in Scandinavian countries, environmental taxes are treated mainly as a form of governmental support for the protection of the environment.

The effectiveness of green transformation is also examined from the perspective of legal conditions (the legal perspective). Fankhauser and Hutter (2016) analysed the relations between the number of the approved decrees related to energy regulation and the laws on environmental protection and the achieved results regarding climate policies. They pointed out that the actual number of regulations implemented in this area does not strengthen the climate policy, whose effectiveness is mostly demonstrated by the quality of the introduced legislation. This perspective, even though relevant in the process of green transformation, is in fact difficult to verify empirically—above all due to the scarcity of data available in official statistics, and to the qualitative character of this kind of conditions.

Green transformation is also defined (e.g. Sattich 2014) in the context of change occurring in the energy sector, mostly regarding its economic effects and technological developments (the economic and technological perspective). According to Stephanides et al. (2019), micro-scale investment in green technologies constitutes a factor which significantly accelerates the green transformation of the entire economy. Heiskanen (2018) point out that company investment in technological solutions dedicated to green economy can also speed up green transformation on the macroeconomic level. The level of technological development attained by the analysed economy also impacts on the pace and course of the green transformation process. The majority of studies examining the conditions for green economy describe it as a key factor (see Bramstoft et al. 2018; Zillman et al. 2018; Sarkar et al. 2019). In particular, they draw attention to the availability and effectiveness of the existing technologies, and to the level of development of new technologies allowing to produce energy from renewable (Magnani and Osti 2016). However, it is worth stressing that despite indicating it as the key factor, the success of the process of transformation is influenced by a whole range of other conditions, including legal and institutional, the research and development attainments, as well as the competitiveness of the economy.

The evaluation of the level and directions of green transformation changes is also significantly affected by factors describing environmental aspects (the environmental perspective), which include information about the emission of pollutants and their impact on the quality of life. The research carried out by Liu and Matsushima (2019) suggests the existence of a notable dependency between the factors describing the effectiveness of obtaining energy and the quality of life, with crucial importance given mostly to the emission of pollutants and of greenhouse gases. Burke et al. (2022) even set the points when the growing cost of the conventional forms of energy will become so high that it will justify bearing the cost of transition towards the renewable sources of energy. The quality of life in the conducted research was mainly linked with economic security.

The pace and the course of the green transformation process also depend on the so-called contextual factors, not connected directly but nevertheless closely linked to it. In the subject literature with green transformation (see: Barbier 2020), these include: GNP per capita, population density, levels of employment and education.

It is also worth noting the different scopes of defining the term 'green transformation'. Narrowly defined, it means a transition from a system of energy based on non-renewable sources of energy, to a system based on renewable ones (Cleveland and Morris 2005). Its broad understanding (cf. Sung and Park 2018), describes it as "a collective, complex and long-term process comprising multiple actors for social changes, involving far-reaching societal changes". A similar interpretation was given by Akermi et al. (2017). According to these authors, "the green energy transition must be understood as a complex process requiring the involvement of many actors". They pointed out that civil society stakeholders' opinions on the utilisation of energy sources have more and more significant impact on central government's actions. It should be stressed here that many studies simply ignore the role of decisions made by citizens in the process of green transformation. There are very few studies which analyse the impact of public opinion on the speed of green transformation. One can mention here the following publications: Tsagarakis et al. (2011), Zahari and Esa (2016), who state that the level of concern about the environment demonstrated by society and the social awareness play a key role in green energy transformation. According to Meadowcroft et al. (2019), the influence exercised by public opinion exceeds that resulting from government actions. Igos et al. (2015) highlight the need to educate society and to demonstrate that actions taken by individuals together constitute a comprehensive system of supporting green transformation. However, some authors (e.g. Sung and Park 2018) claim that the role of society should not be overestimated, and that decisions made by individuals affect green transformation only indirectly. Pachauri and Jiang (2008) suggest that public opinion may only indirectly contribute to the speeding up of green transformation through generating debate and prompting societal innovation.

The kind of approach applied in defining the term 'green transformation' (either wide or narrow) and the choice of conditions for that process (factors of the transformation) are of vital importance for the organisation of empirical research. Unfortunately, official statistics lack data allowing to include the engagement of society in the green transformation process. Data of this kind require conducting empirical research, for example questionnaire surveys among residents. The availability of data useful in constructing a model/s of green transformation, mainly because of the relatively short duration of research carried out internationally, is absolutely crucial. This will also affect the considerations presented further on in this chapter.

3 Research Methodology

3.1 Statistical Data

The basis for constructing a model/s of green transformation in EU countries was provided by official statistical data published by: Eurostat, the World Bank, and the OECD. Table 2 presents a list of indicators used in this study, divided into five groups. The first group comprises variables which directly describe aspects linked with the use of energy. The next three groups include indicators referring to the following conditions: institutional (group II), economic and technological (group III), and environmental (group IV). In line with the information presented in the literature review, the results obtained in EU countries were analysed also considering the so-called contextual indicators (group V) which, although not connected directly with the process of green transformation, yet have a significant influence on its course. In total, 32 indicators were analysed: 7 from group I, 7 from group II, 4 from group III, 8 from group IV, as well as 6 contextual indicators from group V.

The majority of the analysed indicators was highly diversified. The coefficients of variability for 30 out of the 32 utilised to examine the diagnostic features were at the level from 16.50% ($X_{2.10D}$—energy related tax revenue, % total environmental tax revenue) up to 222.57% ($X_{3.17S}$—development of environment-related technologies, % inventions worldwide). Only for two contextual indicators, namely $X_{5.28S}$—life expectancy at birth and $X_{5.31S}$—employment rate, from 20 to 64 years, as percentage of the total population, the coefficients of variability remained below 15% and were at 3.20% and 6.69%, respectively. More than half of the analysed indicators showed right-sided asymmetry, which means that most of the studied countries obtained results below the average. For the indicators marked as destimulants this is a favourable situation. Left-sided asymmetry was demonstrated by 8 out of the 32 examined indicators, which means that most of the studied countries obtained results above the average. This is a situation favourable for indicators marked as stimulants—it confirms a higher level of development in the majority of the examined countries.

3.2 Statistical Method

The procedure of building models of green transformation implemented in this chapter was based on three stages of research. The first stage, using the indicators divided into five groups, involved building rankings presenting the level of development of EU countries (and the United Kingdom). The analyses were carried out separately for each group. At this stage of the study rankings were built utilising the method of relative taxonomy, whose main advantage was the possibility of comparing each indicator set for a specific EU country and the UK with the indicators attained by all the other countries. Relative taxonomy allows to compare the obtained results

Table 2 The base of indicators

Description	Symbol	Descriptive statistics		
		$\bar{x}$	V_s (%)	Asl
I. Energy				
Energy intensity, TPES per capita	$X_{1.1S}$	3.19	37.57	1.17
Total primary energy supply, tonnes of oil equivalent (toe), millions	$X_{1.2D}$	56.46	131.78	2.02
Renewable energy supply, % TPES	$X_{1.3S}$	17.66	58.52	1.08
Renewable electricity, % total electricity generation	$X_{1.4S}$	37.88	56.16	0.56
Energy consumption in agriculture, % total energy consumption	$X_{1.5D}$	2.65	50.98	1.15
Energy consumption in industry, % total energy consumption	$X_{1.6D}$	22.93	29.48	0.87
Energy consumption in transport, % total energy consumption	$X_{1.7D}$	30.94	26.70	0.90
II. Institutional perspective (institutional conditions)				
Environmentally related taxes, % GDP	$X_{2.8D}$	2.47	33.08	−0.13
Environmentally related taxes, % total tax revenue	$X_{2.9D}$	6.87	32.64	−0.26
Energy related tax revenue, % total environmental tax revenue	$X_{2.10D}$	75.03	16.50	−0.70
Road transport-related tax revenue, % total environmental tax revenue	$X_{2.11D}$	22.82	74.52	1.53
Petrol end-user price, USD per litre	$X_{2.120D}$	2.14	21.52	0.25
Diesel tax, USD per litre	$X_{2.13D}$	0.67	20.27	−0.45
Diesel end-user price, USD per litre	$X_{2.14D}$	2.03	24.57	0.51
III. Economic and technological perspective (economic and technological conditions)				
Development of environment-related technologies, % all technologies	$X_{3.15S}$	10.82	51.61	0.78
Relative advantage in environment-related technology	$X_{3.16S}$	1.19	51.72	0.80
Development of environment-related technologies, % inventions worldwide	$X_{3.17S}$	0.94	222.57	3.95
Development of environment-related technologies, inventions per capita	$X_{3.18S}$	13.49	116.23	1.79
IV. Environmental conditions (environmental perspective)				
Mean population exposure to PM2.5, micrograms per cubic metre	$X_{4.19D}$	12.89	33.55	0.22
Percentage of population exposed to more than 10 μg/m³, %	$X_{4.20D}$	69.93	45.48	−1.17
Mortality from exposure to ambient PM2.5, per 1,000,000 inhabitants	$X_{4.21D}$	401.66	68.15	1.31

(continued)

Table 2 (continued)

Description	Symbol	Descriptive statistics		
Welfare costs of premature mortalities from exposure to ambient PM2.5, GDP equivalent, %	$X_{4.22D}$	4.15	72.42	1.32
Mortality from exposure to lead, per 1,000,000 inhabitants	$X_{4.23D}$	25.06	62.97	0.55
Welfare costs of premature deaths from exposure to lead, GDP equivalent, %	$X_{4.24D}$	0.25	64.70	0.61
Mortality from exposure to lead, per 1,000,000 inhabitants	$X_{4.25D}$	85.51	69.75	1.50
Welfare costs of premature deaths from exposure to lead, GDP equivalent, %	$X_{4.26D}$	0.87	72.49	1.50
V. Contextual indicators				
Real GDP per capita, USD Dollar	$X_{5.27S}$	42,697.57	41.43	2.21
Life expectancy at birth	$X_{5.28S}$	80.39	3.20	−0.82
People at risk of poverty or social exclusion, %	$X_{5.29D}$	21.20	23.94	0.60
Tertiary educational attainment, from 25 to 34 years, %	$X_{5.30D}$	42.79	20.10	−0.07
Employment rate, from 20 to 64 years, % of total population	$X_{5.31S}$	74.61	6.69	−1.03
Purchasing power adjusted GDP per capita, %	$X_{5.32S}$	31,867.86	40.49	2.32

Note $\bar{x}$—mean, V_S—coefficient of variation, A_s—asymmetry, S—stimulants and D—destimulants

in pairs (each country with another one), unlike with most of other known taxonomic methods, e.g. with the average level of the examined phenomenon or the established level of reference (i.e. with pattern and anti-pattern).

The synthetic measure in this method is calculated in several steps. In the first one, the relative ratios for each of EU countries against all the others is calculated based on the following formula (Wydymus 2013):

$$d_{(l/i)j} = x_{lj}/x_{ij}, \quad (1)$$

where: d—relativised values of the indicators $i, l = 1, \ldots, k$—countries' numbers, $i \neq l, j = 1, \ldots, m$—numbers of sub-indicators.

It should be noted that the characteristics, which are stimulants, were analysed. Conversions of counter-interpretation to stimulants (i.e. destimulants) can also be used, e.g. for quotient transformations.

The structure of the individual arrays for each j-index can be presented as follows:

$$D_j = \begin{bmatrix} 1 & d_{(2/1)j} & \ldots & d_{(k/1)j} \\ d_{(1/2)j} & 1 & \ldots & d_{(k/2)j} \\ \ldots & \ldots & 1 & \ldots \\ d_{(1/k)j} & d_{(2/k)j} & \ldots & 1 \end{bmatrix}. \quad (2)$$

Next, based on the array of D_j matrices, objects (in this case—EU countries) were classified taking into account the whole set of diagnostic indicators X used for the analysis as follows:

$$A = \begin{bmatrix} 0 & & \frac{1}{(k-1)} \\ & \ldots & \\ \frac{1}{((k-1))} & & 0 \end{bmatrix}, \quad (3)$$

and products $D_j^* = A \cdot D_j$. The elements on the main diagonal matrix D^* form a three-dimensional matrix W defined for all j indicators and periods t:

$$W = \begin{bmatrix} w_{11} & w_{12} & \ldots & w_{1m} \\ w_{21} & w_{22} & \ldots & w_{2m} \\ \ldots & \ldots & \ldots & \ldots \\ w_{k1} & w_{k2} & \ldots & w_{km} \end{bmatrix}. \quad (4)$$

Finally, the relative synthetic measure of development determined by the W matrix was calculated based on the formula:

$$S_i = \left[\sum 1/w_{ij}\right]/m. \quad (5)$$

This measure is close to 1 and can be interpreted as the relative position of the object relative to all other analysed objects. For objects with a similar level of development, the values generally oscillate around unity. The lower the value of the measure, the better the situation of the object (companies) against the background.

The objects can also be divided into typological classes grouped with similar levels of development. The assignment of objects to the typological groups was carried out as follows:

$$group = \begin{cases} 1 \text{ for } S_i \leq \overline{S}_i - S_{Si} \\ 2 \text{ for } S_i \leq \overline{S}_i \quad \wedge S_i > \overline{S}_i - S_{Si} \\ 3 \text{ for } S_i \leq \overline{S}_i + S_{Si} \wedge S_i > \overline{S}_i \\ 4 \text{ for } S_i > \overline{S}_i + S_{Si} \end{cases} \quad (6)$$

where $\overline{S}_{it}$ means the average value determined on the basis of the relative synthetic measure of development, while S_{Sit} their standard deviation.

The first class includes the best objects with the lowest values of the relative synthetic measures, and the fourth class the worst with the highest values.

At the second stage of the research procedure, in order to examine the relationship between the main area of green energy and other examined areas (II–IV), as well as a synthetic measure calculated based on contextual indicators, the following correlation coefficients were used: Pearson's r for the values of synthetic measures, and Kendall's tau for selected items in the constructed rankings.

The final stage of the study attempted to describe the identified regularities which constitute the basis for recognising the models of green transformation in EU countries. Such models allow to demonstrate the differences in approach to the green transformation process in these countries, in particular in the scope of energy transformation.

4 Study Results

Table 3 presents the results of the first stage of the study, showing the values of taxonomic measures of development (S_i), which were set separately for each analysed group, and the positions of the studied countries in the constructed rankings (Formula 5).

The information presented in Table 3 suggests that the results obtained by these countries within the distinguished groups vary significantly. In the ranking regarding the use of energy in the economy, including renewable sources of energy, the first three places were taken by Lithuania, Austria and Sweden. Although these countries were classified in the top positions in the created ranking, for some of the analysed countries the indicators were 'worse' (lower for stimulants and higher for destimulants) than average in the group. This applied to:

(a) two indicators for Lithuania: $X_{1.1S}$ (energy intensity, TPES per capita), and $X_{1.7D}$ (energy consumption in transport, % total energy consumption),
(b) two indicators for Austria: $X_{1.6D}$ (energy consumption in industry, % total energy consumption), and $X_{1.7D}$ (energy consumption in transport, % total energy consumption),
(c) one indicator for Sweden: $X_{1.6D}$ (energy consumption in industry, % total energy consumption).

It is also worth noting that none of the three countries was found in the group of the top ten regarding each of the analysed areas. Sweden, which along with Finland is indicated as an example of a country able to permanently separate economic growth from the negative pressure placed on the natural environment, obtained the decidedly low position (24) in respect of the second analysed area of the institutional perspective (institutional conditions), mainly due to very high environmental taxes. However, on the assumption that such taxes can also imply a significant contribution on the part of large companies to environmental protection, Sweden's position would be definitely much higher. The lowest positions in the ranking were taken by Poland (26), France (27), and Germany (28). These low places resulted from – much higher than for other countries – the values of indicator $X_{1.2D}$ – total primary energy supply (tonnes of oil equivalent (toe), millions), which reached respectively the values of 103.58, 241.40 and 300.83, with an average of 56.46. The classification of EU countries into typological groups referring to this area is shown in Fig. 2, whereas Fig. 3 presents

Table 3 Study results—the results of synthetic measure (S_i) and place

Country	S_1	Place	S_2	Place	S_3	Place	S_4	Place	S_5	Place
Austria	0.725	2	0.929	7	0.327	2	0.881	12	0.859	6
Belgium	1.021	17	0.926	6	0.571	13	1.103	16	0.894	8
Bulgaria	0.986	16	0.936	10	1.304	20	2.340	28	1.415	28
Croatia	0.765	4	1.079	21	4.068	28	1.561	26	1.214	25
Cyprus	1.281	23	0.810	2	1.305	21	1.200	18	1.025	16
Czech Republic	1.106	19	1.153	27	0.555	11	1.214	20	0.960	14
Denmark	0.802	9	1.292	28	0.207	1	0.677	8	0.826	4
Estonia	0.785	7	0.936	9	2.677	27	0.304	4	1.040	18
Finland	0.782	6	1.029	14	0.357	4	0.188	1	0.880	7
France	1.576	27	1.006	13	0.370	5	0.724	9	0.897	9
Germany	1.600	28	1.105	23	0.333	3	0.893	13	0.911	10
Greece	0.876	12	0.960	11	1.348	22	1.577	27	1.221	26
Hungary	1.101	18	0.850	4	1.065	19	1.502	23	1.121	22
Ireland	0.873	11	0.811	3	0.825	16	0.289	3	0.761	2
Italy	1.159	20	1.073	20	0.518	9	1.472	22	1.130	23
Latvia	0.814	10	1.047	15	1.623	24	0.865	11	1.153	24
Lithuania	0.642	1	0.933	8	2.071	25	0.676	7	1.029	17
Luxembourg	0.780	5	1.068	19	0.648	14	0.561	6	0.729	1
Malta	1.384	25	1.062	17	1.574	23	1.529	24	0.936	11
Netherlands	1.359	24	1.135	25	0.482	8	0.798	10	0.815	3
Poland	1.429	26	0.918	5	0.567	12	1.357	21	1.048	19
Portugal	0.891	13	0.659	1	1.060	18	1.031	14	1.061	21
Romania	0.925	15	0.969	12	2.358	26	1.540	25	1.333	27
Slovak Republic	0.917	14	1.098	22	0.912	17	1.132	17	1.054	20
Slovenia	0.793	8	1.062	18	0.702	15	1.046	15	0.937	12
Spain	1.194	22	1.056	16	0.538	10	1.205	19	1.020	15
Sweden	0.739	3	1.108	24	0.370	6	0.276	2	0.846	5
United Kingdom	1.190	21	1.138	26	0.417	7	0.501	5	0.938	13

the values of all the synthetic measures for the three highest and lowest classified countries in respect of the area describing the use of energy in the economy.

The entire presented classification did not show a single country which would obtain top-ten results for the highest ranked countries in the case of each analysed area. The same applies to countries ranked lowest in terms of their use of energy in the economy. This is confirmed by the large diversity of the results obtained in each of the analysed areas. This in turn brought fairly low evaluations for the coefficients of Pearson's correlation r between Area I (energy) and all the other examined areas

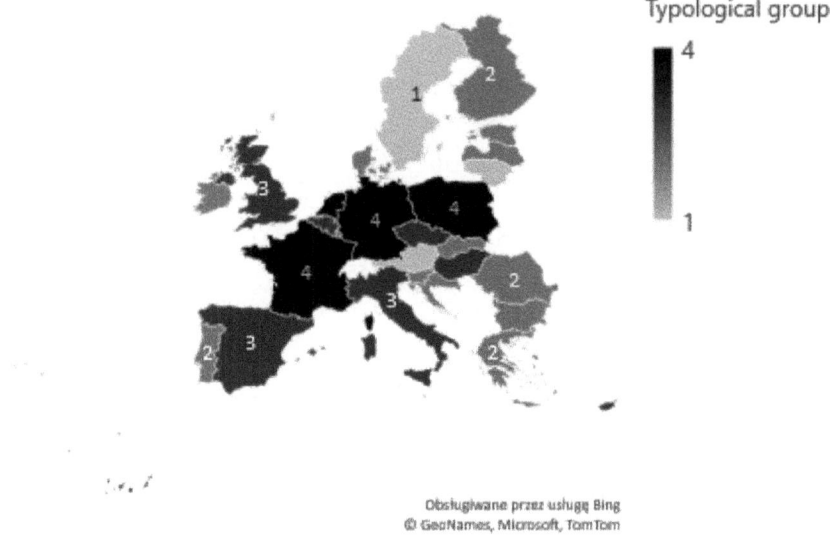

Fig. 2 Division of EU countries into typological groups

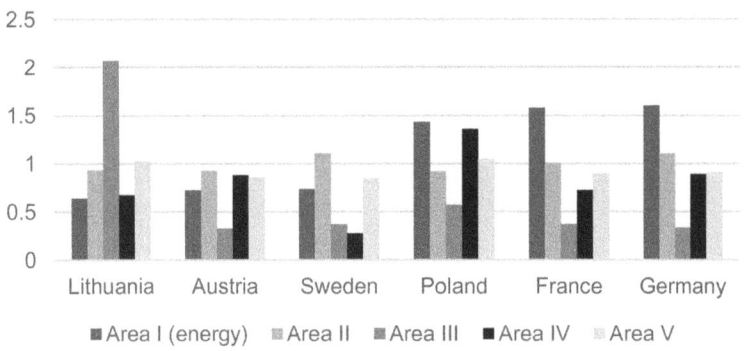

Fig. 3 Values of synthetic measures for the highest and lowest classified countries in Area I (energy)

determined for the values of synthetic measures (Table 4), and Kendall's tau for the positions in the constructed rankings (Table 5).

One can indicate the existence of relatively strong dependencies only for the following: S_3 (economic and technological perspective) and S_5 (contextual indicators), as well as S_4 (environmental conditions) and S_5 (contextual indicators). Such a significant differentiation of the obtained results means that it is actually impossible to indicate one or even a few dominating models of green transformation.

Only the results obtained by two countries (Austria and Ireland) in each of the examined areas allowed to classify both of them into the first two typological groups which comprise countries with the results of synthetic measures above the group

Table 4 Evaluation of coefficients for Pearson's r

Area	S_1	S_2	S_3	S_4	S_5
S_1	1.000	0.071	−0.327	0.234	−0.069
S_2	0.071	1.000	−0.202	−0.143	−0.245
S_3	−0.327	−0.202	1.000	0.290	0.578
S_4	0.234	−0.143	0.290	1.000	0.761
S_5	−0.069	−0.245	0.578	0.761	1.000

Table 5 Evaluation of coefficients for Kendall's tau

Area	S_1	S_2	S_3	S_4	S_5
S_1	1.000	0.005	−0.175	0.233	0.032
S_2	0.005	1.000	−0.259	−0.106	−0.169
S_3	−0.175	−0.259	1.000	0.307	0.466
S_4	0.233	−0.106	0.307	1.000	0.524
S_5	0.032	−0.169	0.466	0.524	1.000

average. The results obtained by four other countries, namely Spain, the Slovak Republic, Slovenia and Belgium were close (the differences in the classification did not exceed one group) and allowed for including them in the second (results above the mean) or third group (the results below the mean). The synthetic values for the other examined countries were far more diversified, which led to classifying them into all the distinguished typological groups (from I to IV). The division into typological groups (Formula 6) is presented in Table 6.

The subject literature also draws attention to the fact that the results obtained by world economies in various areas, such as innovation, sustainable development, quality of life, etc., are highly dependent on society's wealth, however this kind of hypothesis was not confirmed in the study. The evaluation of Pearson's coefficient r set for the variable describing real GDP per capita (USD dollar, $X_{5.27S}$), and of the synthetic measure for Area I, describing use of energy in the economy (S_1), calculated based on the results of all the studied countries, was around −0.08, which in fact means a lack of correlation between these areas. However, its value was decidedly higher for the 14 wealthiest countries, and amounted to −0.41. This result suggests the existence of a moderate dependence—in line with a growing GNP per capita, the value of the synthetic measure in the first analysed area is decreasing, which in the case of relative taxonomy means improvement.

The subject literature also draws attention to the fact that the results obtained by world economies in various areas, such as innovation, sustainable development, quality of life, etc., are highly dependent on society's wealth, however this kind of hypothesis was not confirmed in the study. The evaluation of Pearson's coefficient r set for the variable describing real GDP per capita (USD dollar, $X_{5.27S}$), and of the synthetic measure for Area I, describing use of energy in the economy (S_1), calculated

Table 6 The results of division into typological groups

Country	Area				
	I	II	III	IV	V
Austria	1	2	2	2	2
Belgium	3	2	2	3	2
Bulgaria	2	2	3	4	4
Croatia	2	3	4	4	4
Cyprus	3	1	3	3	3
Czech Republic	3	4	2	3	2
Denmark	2	4	1	2	1
Estonia	2	2	4	1	3
Finland	2	3	2	1	2
France	4	3	2	2	2
Germany	4	3	2	2	2
Greece	2	2	3	4	4
Hungary	3	1	3	4	4
Ireland	2	1	2	1	1
Italy	3	3	2	4	4
Latvia	2	3	4	2	4
Lithuania	1	2	4	2	3
Luxembourg	2	3	2	2	1
Malta	4	3	4	4	2
Netherlands	4	4	2	2	1
Poland	4	2	2	3	3
Portugal	2	1	3	3	3
Romania	2	2	4	4	4
Slovak Republic	2	3	3	3	3
Slovenia	2	3	2	3	2
Spain	3	3	2	3	3
Sweden	1	3	2	1	2
United Kingdom	3	4	2	1	2

based on the results of all the studied countries, was around -0.08, which in fact means a lack of correlation between these areas. However, its value was decidedly higher for the 14 wealthiest countries, and amounted to -0.41. This result suggests the existence of a moderate dependence—in line with a growing GNP per capita, the value of the synthetic measure in the first analysed area is decreasing, which in the case of relative taxonomy means improvement.

5 Summary

The analyses presented in this chapter suggest that it is practically impossible to construct one or even a few models of green transformation for countries in the EU (and United Kingdom). The results of individual countries were strongly diversified. Only two of the studied countries, i.e. Austria and Ireland, obtained results which allowed to classify them into the first two typological groups in the individual areas; the results of other countries were definitely more varied.

Currently, EU countries are at different stages of their development and approach to using green energy in their economies. In coming years, this situation will bring many difficulties in attempting a systemic approach to formulating directives regarding the transition from a system using non-renewable sources of energy to one based on sources that are renewable. There are also differences between the studied countries in terms of the acceptance and the consensus of society to bear the higher cost of this kind of transformation. Nowadays there is a growing interest, not merely from governments but also entire communities, in the conditions of transition towards an economy based on renewable sources of energy. The debate in this scope, not just in the context of protecting the natural environment, as well as in terms of safe access to energy—in the situation of the war in Ukraine—is being undertaken by all European governments. Its effect should be the acceleration of this process accompanied by the development of new technologies for obtaining energy from renewable sources.

References

Akermi R, Hachana ST, Triki A (2017) Conceptualizing civil society attitudes towards the promotion of renewable energy: a case study from Tunisia. Energy Procedia 141:131–137

Amundsen H, Hermansen EA (2021) Green transformation is a boundary object: an analysis of conceptualisation of transformation in Norwegian primary industries. Environ Plan E: Nat Space 4(3):864–885. https://doi.org/10.1177/2514848620934337

Barbier EB (2020) Greening the post-pandemic recovery in the G20. Environ Resource Econ 76(4):685–703. https://doi.org/10.1007/s10640-020-00437-w

Bjørner TB, Jensen HH (2002) Energy taxes, voluntary agreements and investment subsidies—a micro-panel analysis of the effect on Danish industrial companies' energy demand. Resource Energy Econ 24(3):229–249. https://doi.org/10.1016/S0928-7655(01)00049-5

Borel-Saladin JM, Turok IN (2013) The green economy: incremental change or transformation? Environ Policy Gov 23(4):209–220. https://doi.org/10.1002/eet.1614

Bramstoft R, Alonso AP, Karlsson K, Kofoed-Wiuff A, Münster M (2018) STREAM–an energy scenario modelling tool. Energ Strat Rev 21:62–70

Burke PJ, Beck FJ, Aisbett E, Baldwin KG, Stocks M, Pye J, Venkataraman M, Hunt J, Bai X (2022) Contributing to regional decarbonization: Australia's potential to supply zero-carbon commodities to the Asia-Pacific. Energy 248:123563

Chou J, Mathews JA (2017) Taiwan's green energy transition under way. The Asia Pac J, Jpn Focus 15(21):1–9

Cleveland CJ, Morris, CG (eds) (2005) Dictionary of energy. Elsevier

Crespi F, Mazzanti M, Managi S (2016) Green growth, eco-innovation and sustainable transitions. Environ Econ Policy Stud 18(2):137–141. https://doi.org/10.1007/s10018-016-0141-x

Cui H, Lui Z (2021) Spatial-temporal pattern and influencing factors of the urban green development efficiency in Jing-Jin-Ji Region of China. Pol J Environ Stud 30(2). https://doi.org/10.15244/pjoes/124758

Declich A, Quinti G, Signore P (2020) SME's, energy efficiency, innovation: a reflection on materials and energy transition emerging from a research on SMEs and the practice of energy audit. Matér Tech 108(5–6):505. https://doi.org/10.1051/mattech/2020036

Devine-Wright P, Murphy J (eds) (2007) Energy citizenship: psychological aspects of evolution in sustainable energy technologies. In: Framing the present, shaping the future: contemporary governance of sustainable technologies. Earthscan

Fankhauser P, Hutter M (2016) A universal grid map library: implementation and use case for rough terrain navigation. In: Robot operating system (ROS). Springer, Cham, pp 99–120

Feng Z, Chen W (2018) Environmental regulation, green innovation, and industrial green development: an empirical analysis based on the spatial Durbin model. Sustainability 10(1):223. https://doi.org/10.3390/su10010223

Feola G (2015) Societal transformation in response to global environmental change: a review of emerging concepts. Ambio 44(5):376–390. https://doi.org/10.1007/s13280-014-0582-z

Gea-Bermúdez J, Jensen IG, Münster M, Koivisto M, Kirkerud JG, Chen YK, Ravn H (2021) The role of sector coupling in the green transition: a least-cost energy system development in Northern-central Europe towards 2050. Appl Energy 289:116685. https://doi.org/10.1016/j.apenergy.2021.116685

Heiskanen HE (2018) Towards greener human rights protection: rewriting the environmental case law of the European court of human rights

Hu C, Mao J, Tian M, Wei Y, Guo L, Wang Z (2021) Distance matters: Investigating how geographic proximity to ENGOs triggers green innovation of heavy-polluting firms in China. J Environ Manage 279:111542. https://doi.org/10.1016/j.jenvman.2020.111542

Igos E, Rugani B, Rege S, Benetto E, Drouet L, Zachary DS (2015) Combination of equilibrium models and hybrid life cycle-input–output analysis to predict the environmental impacts of energy policy scenarios. Appl Energy 145:234–245

Lennon B, Dunphy NP, Sanvicente E (2019) Community acceptability and the energy transition: a citizens' perspective. Energ Sustain Soc 9:35. https://doi.org/10.1186/s13705-019-0218-z

Liu B, Matsushima J (2019) Annual changes in energy quality and quality of life: a cross-national study of 29 OECD and 37 non-OECD countries. Energy Rep 5:1354–1364

Magnani N, Osti G (2016) Does civil society matter? Challenges and strategies of grassroots initiatives in Italy's energy transition. Energy Res Soc Sci 13:148–157

Meadowcroft J, Banister D, Holden E, Langhelle O, Linnerud K, Gilpin G (eds) (2019) What next for sustainable development?: our common future at thirty. Edward Elgar Publishing

Pachauri S, Jiang L (2008) The household energy transition in India and China. Energy Policy 36(11):4022–4035

Rosenow J, Kern F, Rogge K (2017) The need for comprehensive and well targeted instrument mixes to stimulate energy transitions: the case of energy efficiency policy. Energy Res Soc Sci 33:95–104. https://doi.org/10.1016/j.erss.2017.09.013

Ruszel M (2017) The role of energy resources in electricity production in the EU up to 2050. Polityka Energetyczna—Energy Policy J 20(3):5–15

Sarkar B, Tayyab M, Kim N, Habib MS (2019) Optimal production delivery policies for supplier and manufacturer in a constrained closed-loop supply chain for returnable transport packaging through metaheuristic approach. Comput Ind Eng 135:987–1003

Sattich T (2014) Germany's energy transition and the European electricity market: mutually beneficial? J Energy Power Eng 8(2)

Schot J, Kanger L, Verbong G (2016) The roles of users in shaping transitions to new energy systems. Nat Energy 1:16054. https://doi.org/10.1038/nenergy.2016.54

Stephanides P, Chalvatzis KJ, Li X, Mantzaris N, Prodromou M, Papapostolou C, Zafirakis D (2019) Public perception of sustainable energy innovation: a case study from Tilos, Greece. Energy Procedia 159:249–254

Sung B, Park SD (2018) Who drives the transition to a renewable-energy economy? Multi-actor perspective on social innovation. Sustainability 10(2):448

Tsagarakis KP, Bounialetou F, Gillas K, Profylienou M, Pollaki A, Zografakis N (2011) Tourists' attitudes for selecting accommodation with investments in renewable energy and energy saving systems. Renew Sustain Energy Rev 15(2):1335–1342

Wang MX, Zhao HH, Cui JX, Fan D, Lv B, Wang G, Zhou GJ (2018) Evaluating green development level of nine cities within the Pearl River Delta, China. J Clean Prod 174:315–323. https://doi.org/10.3390/su13063034

Wydymus S (2013) Rozwój gospodarczy a poziom wynagrodzeń w krajach unii europejskiej–analiza taksonomiczna. Zeszyty Naukowe Uniwersytetu Szczecińskiego 756:632–645

Zahari AR, Esa E (2016) Motivation to adopt renewable energy among generation Y. Procedia Econ Finan 35:444–453

Zillman D, Godden L, Paddock L, Roggenkamp M (eds) (2018) Innovation in energy law and technology: dynamic solutions for energy transitions. Oxford University Press

Ziolo M, Jednak S, Savić G, Kragulj D (2020) Link between energy efficiency and sustainable economic and financial development in OECD countries. Energies 13(22):5898

The Study of Disproportions in the Area of Green Energy in EU Countries

Maciej Oesterreich and Katarzyna Wawrzyniak

Abstract The chapter provides a quantitative analysis of the disparity between EU countries in the use of renewable energy. For this purpose, the synthetic measures determined by the TOPIS method and the structures similarity coefficient were used. The statistical data used in the analysis came from the EUROSTAT database and contained information from 2016 to 2020. The conducted analyses confirmed the existence of differences both in the level of use of energy from renewable sources and the structure of its production. The changes in energy balances necessitated by Directive 2009/28/EC of the European Parliament and the Council and the goals stipulated in the Paris Agreement also unfolded differently in the countries studied. They resulted in the majority of members achieving the prescribed limits. The highest share of energy from renewable sources was recorded among northern European and Baltic countries. In contrast, it developed the fastest in Central and Eastern Europe. The lowest level of use of green energy was detected among the Balkan countries and in France.

1 Introduction

The diversification of energy sources by individual countries of the European Union, with particular emphasis on renewable sources, is currently essential for environmental reasons and the need to become independent from traditional energy sources (oil, gas, coal) imported from outside the country. In order to answer the question regarding the current status of green energy in EU countries, the disproportions in production volume and consumption of energy from renewable sources (RES) in these countries were examined.

M. Oesterreich (✉) · K. Wawrzyniak
Faculty of Economics, West Pomeranian University of Technology in Szczecin, Szczecin, Poland
e-mail: maciej.oesterreich@zut.edu.pl

K. Wawrzyniak
e-mail: katarzyna.wawrzyniak@zut.edu.pl

© The Author(s), under exclusive license to Springer Nature Switzerland AG 2023
I. Bąk and K. Cheba (eds.), *Green Energy*, Green Energy and Technology,
https://doi.org/10.1007/978-3-031-12531-7_8

Fourteen indicators were used in the study of green energy disparities. In order to select them, the question was asked: what aspects related to green energy can differentiate the analysed countries? In order to find an answer to this question, a review of literature on the differentiation of countries and regions in terms of production and consumption of energy from renewable sources and studies (reports) of public statistics was conducted.[1] On this basis, it was assumed that significant differences (disproportions) in the field of green energy between EU countries might concern the following aspects:

- the current (2020) level of production and consumption of energy from renewable sources,
- the transformation rate of energy sources (from traditional to renewable energy sources) over the last five years (2016–2020),
- the degree of implementation of EU directives stipulating target values that EU countries should achieve in 2020 for selected indicators.

The selected indicators were divided into three groups. The first group included indicators characterising the level of production and consumption of energy from renewable sources in the EU countries in 2020[2]:

- X_1—the share of energy from renewable sources in total primary energy (%),
- X_2—the share of electricity from renewable sources in gross energy production (%),
- X_3—the share of electricity from renewable sources in the gross final energy consumption (%),
- X_4—the share of energy from renewable sources in gross final energy consumption (%),
- X_5—the share of energy from renewable sources in gross final energy consumption in transport (%),
- X_6—share of energy from renewable sources in gross final energy consumption in heating and cooling (%).

Indicators in the second group that characterise the rate of the green energy transformation were calculated as the difference between the values of indicators from the first group in 2020 and 2016.[3] The choice of 2016 as the base period was deliberate, as it was this year in which the green energy transition process accelerated significantly, mainly due to:

---

[1] For example, in Poland, the STATISTICS POLAND publishes data and results of analyses in the field of the acquisition and consumption of energy from renewable sources annually since 2006 in studies entitled: Energy from renewable sources in 2006, 2007, …, 2020.

[2] The names of the indicators in this group are consistent with the names in the EUROSTAT database.

[3] In establishing these indicators, it was decided to apply an absolute increment for the compared shares, since the calculation of the average absolute gain was not justified because, in most EU countries, the values of the first group indicators (X_1–X_6) did not show a one-way change over the years 2016–2020.

- the signing of the Paris Agreement (UNFCCC 2022) in December 2015, the main objective of which is to limit the average temperature increase on Earth,
- approaching the deadline (the end of 2020) by which individual EU countries should achieve the objectives set out in Directive 2009/28/EC of the European Parliament and of the Council of 23 April 2009 (EU 2009) on the promotion of the use of energy from renewable sources.

This group also includes six indicators with the following symbols and names:

- X_7—change in the share of energy from renewable sources in the total primary energy in 2020 compared to 2016 (percentage point (p.p.)),
- X_8—change in the share of electricity from renewable sources in gross energy production in 2020 compared to 2016 (p.p.),
- X_9—change in the share of electricity from renewable sources in gross final energy consumption in 2020 compared to 2016 (p.p.),
- X_{10}—change in the share of energy from renewable sources in the gross final energy consumption in 2020 compared to 2016 (p.p.),
- X_{11}—change in the share of energy from renewable sources in gross final energy consumption in transport in 2020 compared to 2016 (p.p.),
- X_{12}—change in the share from renewable sources in gross final energy consumption in heating and cooling in 2020 compared to 2016 (p.p.).

The last group included two indicators, which were calculated as the difference between the values of these indicators in individual Union countries in 2020 and the targets that Member States should achieve in 2020 following the previously mentioned Directive 2009/28/EC of the European Parliament and of the Council of 23 April 2009 (EU 2009). The names and symbols of these indicators are as follows:

- X_{13}—the degree of implementation (exceeding, reaching or not reaching) of the target share of energy from renewable sources in gross final energy consumption (p.p.),[4]
- X_{14}—the degree of implementation (exceeding, reaching or not reaching) of the target share of energy from renewable sources in the gross final consumption of energy in transport (p.p.).[5]

Based on the above-mentioned indicators, the disproportions in green energy in EU countries were characterised in each area separately, and in total, i.e., all fourteen indicators were considered simultaneously. Thanks to this approach, it was possible to obtain an answer to the question of which aspects and to what extent they influence the level of total disproportions in particular EU countries.

---

[4] For the share of energy from renewable sources in gross final consumption of energy, the 2020 national targets are set out in Annex 1 to Directive 2009/28/EC of the European Parliament and of the Council of 23 April 2009 on the promotion of the use of energy from renewable sources (EU 2009).

[5] For the share of energy from renewable sources in gross final consumption of energy in transport, the target value in 2020 is 10% and is the same for all Member States (Directive 2009/28/EC of the European Parliament and of the Council of 23 April 2009, …, point 8).

In addition, it was decided to compare the structure of renewable energy sources by means in individual EU countries with the average structure, i.e., the structure for EU countries in 2020. Within this structure, the following renewable energy carriers were distinguished: wind, hydro (with the tide, wave, ocean), solar (thermal and photovoltaic), biofuels (solid and liquid), biogases, renewable municipal waste, geothermal and ambient heat (heat pumps). This aspect of the research allowed the authors to identify the specificity of individual countries due to the method of obtaining energy from renewable sources. It thus helped explain the reasons for possible disproportions in green energy.

Due to the fact that the survey was conducted based on the most up-to-date data on the proposed indicators available in the EUROSTAT database (2022), i.e., for 2020, the surveyed population was limited to 27 countries (EU-27), as the data for 2020 were no longer updated for the United Kingdom, which left the EU in 2021.

The TOPSIS method was used to study the disproportions in green energy (in total and in separate areas) in EU countries. In contrast, the structural similarity coefficient was used to study the similarity of structures.

2 Diversification of Countries and Regions in Terms of Production and Consumption of Energy from Renewable Sources—Literature Review

Disparities in both consumption and production of energy from renewable sources result primarily from the physical (geographical) conditions of the regions. Belev (2021) assessed the potential of the regions of Bulgaria to use solar and wind power to generate electricity regarding the minimum technical requirements. He showed that the main barriers to the development of energy sources based on these resources are:

- natural barriers—related to the occurrence of habitats and the preservation of animal and plant species,
- economic—related to agricultural activity and insufficient degree of infrastructure development.

Similar papers related to the technical aspects of the use of energy from renewable sources can be found in the reports: of the Polish Economic Institute (Juszczak and Maj 2020) in relation to Poland and IRENA (2020) in relation to African countries.

In turn, the Sweco report (2019) indicates the following barriers that affect the diversity in the development of energy based on renewable sources in individual countries:

- the availability of fossil fuels—the more readily available they are, the slower the pace of energy transition (Viñuales 2021; AmCham 2022).
- the level of costs, including in particular costs related to energy storage—is currently the biggest drawback of renewable energy sources, whose maximum

level of generated capacity is not constant and depends mainly on weather conditions (CEF 2022).
- the impact of energy transformation, and above all, its costs on the economy and the standard of living of societies (Henning and Palzer 2015; Changyong et al. 2021).
- current government policies do not always support the development of alternative energy sources (De Laurentis and Pearson 2021).
- "energy density", i.e. the amount of land required to generate 1 MW of energy—for example, to generate similar power to that of coal when using solar energy, it is necessary to use more than three times as much land, and in the case of wind almost six times as much (Layton 2008).

An interesting discussion of the impact of income inequality in societies on the share of renewable energy in the energy balance can be found in (McGee and Greiner 2019). The authors examined the condition of 174 economies in the years 1990–2014 and showed that the greater the income inequality, the lower the consumption of energy from renewable sources. As indicated by the authors, such regularity is a consequence of the high level of expenditure necessary to implement investments in green energy. Similar conclusions with regard to sub-Saharan Africa can be found in the paper by (Asongu and Odhiambo 2020) and with regard to 22 OECD countries and China in the paper of (Churchill et al. 2021).

In their paper (Yao et al. 2020), the authors presented the results of an extensive study of inequalities in fossil and renewable energy production and consumption for 57 countries around the world. They used the Theil Inequality Index and applied Generalised Methods of Moments to estimate model parameters. The authors found significant inequalities in energy consumption from renewable sources in the regions of Europe and Central Asia, as well as East Asia and the Pacific while pointing out that their scale decreased over time. At the same time, they found that inequalities in both energy production and consumption negatively affected the environment.

Similar results are presented in the research of Sinha (2017), in which 28 OECD countries were surveyed. In the conclusions of this research, the author emphasised that the energy transition is very much influenced by the institutions (governments) of individual countries (see also Sharvini et al. 2018 and Uzar 2020). He also stated that the unevenness of this transformation across regions should be explained by "disproportionate industrial development and the lack of technology diffusion".

Important findings from a study of renewable energy consumption in households of EU countries were presented by Piekut (2021). It turned out that a higher share of consumption of this energy was recorded in the so-called "poorer part of Europe", i.e., mainly in the countries of the former Eastern Bloc, which were admitted to the EU in 2004 and later. In contrast, the share of renewable sources in final energy consumption was lower in countries with a higher standard of living (Ireland, Luxembourg, United Kingdom, the Netherlands, Belgium, Sweden, and Germany). According to the author, this is mainly due to the larger dwellings and the additional technical infrastructure that increases energy consumption in developed countries. The author's research also shows that the EU leaders in terms of the level of energy consumption from renewable sources in absolute terms are France, Germany and Italy. These countries are also leaders in terms of the pace of the changes from 2004 to 2019. The publication emphasises that the rate of change of renewable energy shares in total energy production in Western European and Scandinavian countries is influenced by governmental actions supporting the development of green energy and consisting of fiscal support, investment subsidies and appropriate education in this respect.

3 The Description of the Methods Used in the Study

3.1 TOPSIS Method

The idea of the TOPSIS method, which belongs to the methods of multidimensional statistical analysis or, more broadly, to multicriteria decision-making methods, is based on the distances of the tested object, described by the values of the set of diagnostic features, from the pattern (a positive ideal object) and the anti-pattern (a negative ideal object). This method assumes that the optimum object for a given diagnostic feature should be characterised by the smallest possible distance from the pattern and the largest possible distance from the anti-pattern. Such an assumption is, at the same time, an advantage of the TOPSIS method because it enables such an evaluation of the object under study from the point of view of many diagnostic features that a bad evaluation of the object from the point of view of one of the features can be compensated by a good evaluation from the point of view of another feature. Based on distances from the pattern and anti-pattern determined for each characteristic, a synthetic measure is calculated, the values of which enable to order and group the studied objects with regard to the level of the studied phenomenon. A detailed description of TOPSIS methods can be found in many publications (e.g., Hwang and Yoon 1981; Pavić and Novoselac 2013; Łatuszyńska 2013; Kozera et al. 2017; Zulqarnain et al. 2020; Bąk et al. 2022). The procedure used in this study will be presented below (one of the steps of the method, i.e., assigning weights to the diagnostic features, has been omitted as the study assumes that all selected indicators are equally important).

The starting point was a matrix with the values of indicators adopted in the study of disproportions in the field of green energy for EU countries in 2020 in the form of:

$$\mathbf{X} = [x_{ij}]_{n \times m} \quad (1)$$

where: x_{ij}—the value of the j-th indicator in the i-th EU country in 2020, $i = 1, 2, \ldots, n$; $n = 27$ $j = 1, 2, \ldots, m$; $m = 14$.

In order to achieve comparability of the indicators, they have been standardised using a multiplying transformation according to the formula:

$$z_{ij} = \frac{x_{ij}}{\sqrt{\sum_{i=1}^{n} x_{ij}^2}} \tag{2}$$

and a matrix of normalised values was obtained:

$$\mathbf{Z} = [z_{ij}]_{n \times m} \tag{3}$$

Then, on the basis of the matrix $\mathbf{Z}$, the coordinates of the pattern (z_j^+—positive ideal object) and the anti-pattern (z_j^-—negative ideal object) for each indicator were determined. Since all indicators used in the study are stimulants (the higher the value of the feature, the higher the level of the studied phenomenon), these coordinates were determined according to the following formulas:

$$z_j^+ = \max_i z_{ij} \tag{4}$$

$$z_j^- = \min_i z_{ij} \tag{5}$$

where: z_j^+—j-th coordinate of the pattern (positive ideal object), z_j^-—j-th coordinate of the anti-pattern (negative ideal object).

Knowing the coordinates of the pattern and the anti-pattern for individual indicators, the distances of the i-th object (EU country) from the pattern and the anti-pattern were calculated:

$$d_i^+ = \sqrt{\sum_{j=1}^{m} (z_{ij} - z_j^+)^2} \tag{6}$$

$$d_i^- = \sqrt{\sum_{j=1}^{m} (z_{ij} - z_j^-)^2} \tag{7}$$

where: d_i^+—distance of the i-th object from the common pattern (positive ideal object), d_i^-—distance of the i-th object from the common anti-pattern (negative ideal object).

Then, based on the determined distances d_i^+ and d_i^- the values of the synthetic measure were calculated, for each tested object, in accordance with the formula:

$$z_i = \frac{d_i^-}{d_i^- + d_i^+} \tag{8}$$

where z_i—value of the synthetic measure for the i-th object (EU country).

A synthetic measure z_i takes the values in the range [0; 1]. The values of this measure close to 1 indicate a high level of the studied phenomenon for a given object, while close to 0 indicates a low level. By using the values of the synthetic measure, it is possible to organise the examined objects from the best to the worst from the point of view of the studied phenomenon and divide them into four typological groups according to the formulas:

$$\text{group 1(best)} : z_i \geq \bar{z} + S(z) \tag{9}$$

$$\text{group 2} : \bar{z} \leq z_i < \bar{z} + S(z) \tag{10}$$

$$\text{group 3} : \bar{z} - S(z) \leq z_i < \bar{z} \tag{11}$$

$$\text{group 4 (theworst)} : z_i < \bar{z} - S(z) \tag{12}$$

where: $\bar{z}$—arithmetic mean of synthetic measure; $S(z)$—standard deviation of synthetic measure.

The presented procedure refers to the study of disproportions in green energy in EU countries altogether, i.e., including all proposed indicators simultaneously. However, in the case of examining disparities in the highlighted areas (current level, rate of change, degree of implementation of target values) from the matrix, **Z** we select respectively sub-matrices $\mathbf{Z}_{s(27\times 6)}$, $\mathbf{Z}_{d(27\times 6)}$, $\mathbf{Z}_{n(27\times 2)}$ and appropriate coordinates of the pattern and anti-pattern for indicators from individual groups. Next, the synthetic measures for a given area, namely z_{si}, z_{di} and z_{ni} respectively, are calculated (according to Formulas 6–8) for the countries under study. In the next step, the countries are ordered and grouped (according to Formulas 9–12) for each area separately.

3.2 Structures Similarity Coefficient

The study of similarity of structures begins with the determination of a structural set, i.e., a set of features $(X_1, X_2, \ldots, X_m)$ through which the examined objects (EU countries) were described and constituting a structure created based on a uniform qualitative or quantitative criterion. The features constituting the structural set must simultaneously meet two conditions (Młodak 2006, pp. 52–53):

(1) normalisation condition, i.e., $x_{ij} \in [0; 1]$,
(2) unit sum condition, i.e.

$$\sum_{j=1}^{m} x_{ij} = 1 \tag{13}$$

where: x_{ij}—share of the j-th feature in the structure of the i-th object, $i = 1, 2, \ldots, n$; n—number of features, $j = 1, 2, \ldots, m$; m—number of features in the structural set.

Considering the above, the formula for the structures similarity coefficient of the i-th and k-th structures can be written as follows (Roszkowska 2020):

$$P_{ik} = \sum_{j=1}^{m} \min(x_{ij}, x_{kj}) \qquad (14)$$

The structures similarity coefficient is a standardised measure in the range [0; 1]. High values of this measure (close to 1) indicate the similarity of the compared objects, and low values (close to 0) indicate the lack of such similarity.

The similarity coefficient of structures calculated according to Formula (14) is also used in the construction of the measure (distance) of the non-similarity of structures and is its complement to unity, i.e. (Chomątowski and Sokołowski 1978; Markowska et al. 2019):

$$d_{ik} = 1 - \sum_{j=1}^{m} \min(x_{ij}, x_{kj}) \qquad (15)$$

It is also worth mentioning that the distance calculated with Formula (15) is equal to the Bray–Curtis Index, included among the measures of object dissimilarity (Sarker and Islam 1999), which for the structured set is calculated according to the formula (Młodak 2006, p. 54):

$$CZ_{ik}^* = \frac{\sum_{j=1}^{m} |x_{ij} - x_{kj}|}{2} \qquad (16)$$

In this study, the structural set consists of eight features $(X_1, X_2, \ldots, X_8)$, which were defined as the shares of renewable energy carriers (wind, hydro, solar, biofuels, biogases, renewable municipal waste, geothermal, heat pumps) in the total energy from renewable sources in 2020. These features form the structure of renewable energy production in individual EU countries and for all countries altogether (EU-27). The structures similarity coefficient was calculated for each EU country separately. Each time, the structure in a given country was compared with the average structure, i.e., the structure for the EU-27. Therefore, in the Formula (14), x_{ij} refers to the share of j-th feature in the structure of i-th Union country, while x_{kj}—the share of the j-th feature in the average structure.

4 Results of the Research on Green Energy Disparities in EU Countries

4.1 Characteristics of the Statistical Material

The study of the disproportions in the field of green energy in the EU countries began with determining basic descriptive parameters characterising the distribution of individual indicators. The values of these parameters are given in Table 1.

From the point of view of examining the disproportions, i.e., the uneven distribution of indicator values in EU countries, the most important is the interpretation of the coefficient of variation, the measure of asymmetry and the difference between the maximum and minimum value of a given indicator. On this basis, it can be determined to what extent the EU countries differ from each other due to the values of individual indicators and whether, among the surveyed countries, countries with index values above or below the average prevail. The greater the variation and differences between the extreme values of indicators, the greater the disproportions from the point of view of a given indicator in the studied countries. In the case of asymmetry—because all

Table 1 Basic descriptive parameters for the tested indicators (diagnostic features)

Feature	Mean ($\bar{x}$)	Minimum (x_{min})	Maximum (x_{max})	Standard deviation ($S(x)$)	Coefficient of variation $\left(V_s = \frac{S(x)}{\bar{x}} \cdot 100\%\right)$	Skewness $A_s = \frac{n \cdot \sum (x_i - \bar{x})^3}{((n-1)\cdot(n-2)\cdot S(x)^3)}$
Indicators characterising the current level of green energy (%)						
X_1	54.636	21.592	100.000	27.371	50.097	0.379
X_2	42.815	11.329	88.562	22.640	52.880	0.396
X_3	35.249	9.489	78.204	19.185	54.429	0.680
X_4	24.360	10.714	60.124	11.471	47.090	1.413
X_5	10.388	5.341	31.854	4.790	46.111	3.593
X_6	30.957	6.264	66.381	17.065	55.123	0.517
Indicators characterising the rate of transformation of energy sources (p.p.)						
X_7	7.275	−2.122	26.928	7.385	101.516	1.158
X_8	6.830	−4.568	34.774	8.308	121.638	1.625
X_9	6.333	0.553	13.857	4.065	64.183	0.488
X_{10}	3.932	−0.554	8.153	2.377	60.457	−0.267
X_{11}	4.146	−0.300	11.734	2.589	62.441	1.093
X_{12}	3.519	−3.422	12.427	3.978	113.035	0.256
Indicators characterising the degree of implementation of EU directives (p.p.)						
X_{13}	2.768	−3.891	11.124	3.281	118.560	1.013
X_{14}	0.388	−4.659	21.854	4.790	1235.224	3.593

the indicators examined are stimulants—a left-hand asymmetry should be considered positive because then, in most countries, the values of a given indicator exceed the average value.

In view of the above information, the interpretation of the results of Table 1 followed. It was found that in the first group of indicators, all indicators are characterised by a fairly strong differentiation ($V_s > 45\%$) and right asymmetry of varying strength, which means a predominance of countries with indicators below the average. The strongest asymmetry occurs in the case of indicator X_5 (the share of energy from renewable sources in gross final energy consumption in transport), in other words, in the vast majority of EU countries, the share of energy from renewable sources is below the average, which in 2020 reached 10.388%. Also, for the indicator X_4 (the share of energy from renewable sources in gross final energy consumption), the asymmetry is very strong, so in most EU countries, energy consumption from renewable sources in 2020 was below 24.36%. Comparing the maximum and minimum values in this group of indicators, one can see that there are countries where the current level of green energy is very high. However, there are also countries where this level is very low. Thus, there are disproportions in the level of production and consumption of energy from renewable sources in the EU countries—the greatest disproportions can be observed for indicator X_1 (the share of energy from renewable sources in the total primary energy), whose value in Malta is 100%, and in Poland, only 21.592%. The smallest disproportions are visible in the indicator X_5 (the share of renewable energy in gross final energy consumption of in transport), what, in combination with the very strong right-hand asymmetry, means that for this indicator—as it turned out after a detailed analysis—only in one country, Sweden, the value was at the level of 31.854%. In the remaining countries, it did not exceed 14%.

In the second group of indicators, characterising the rate of transformation of energy sources in 2016–2020, the diversity was much stronger ($V_s > 60\%$) than for indicators from the first group. It means that the transformation rate of energy sources in individual EU countries deviated quite significantly from the average rate, both in plus (faster-than-average rate of change) and minus (slower-than-average rate of change). The most significant increase was observed for the indicator X_8 (the change in the share of electricity from renewable sources in gross energy production) in Estonia, and amounted to 34.774 p.p., i.e., it was higher than the average increase by more than 28 p.p. For this indicator, the largest decrease in value in the compared years was also observed at the level of 4.568%, and it occurred in Malta. Right-handed asymmetry was dominant in this group of indicators, with very strong asymmetry for three indicators X_7 (change in the share of renewable energy in total primary energy), X_8 (change in the share of renewable electricity in gross energy production) and X_{11} (change in the share of renewable energy in gross final energy consumption in transport) and moderate and weak asymmetry for indicators X_9 (change in the share of renewable electricity in gross final energy consumption) and X_{12} (change in the share of renewable energy in gross final energy consumption in heating and cooling), respectively. It means that for most of the surveyed countries, especially in the case of energy extraction, electricity generation and energy consumption in transport, the rate of transformation of energy sources was slower than the average rate. For the indicator X_{10} (change in the share of energy from renewable sources in the gross final energy consumption), the asymmetry was weak leftward; in other words, there was a slight prevalence of countries with the level of this indicator above the average of 3.932%. The comparison of maximum and minimum values in this group of indicators confirms the earlier observations that among the EU countries, there are countries where the transformation of energy sources is quite dynamic. Unfortunately, there are also countries where the values of the analysed indicators decreased in the compared years.

In the last group, there are only two indicators and the distribution of each of them is characterised by a very strong variation and a very strong right-hand asymmetry. Thus, there is a very strong predominance of those EU countries in which the degree of implementation of the assumed values resulting from the EU directive is below average, and in addition, these averages are at a very low level (for the indicator the X_{13} average is 2.768 p.p.; for the indicator the X_{14} average is 0.388 p.p.). A detailed analysis of the values of these indicators in individual countries made it possible to note that in the case of:

- The indicator X_{13} (the rate of achievement of the target share of renewable energy in gross final energy consumption) only France did not reach the target of 23% in 2020 (3.9% missing); three countries, Belgium, the Netherlands and Slovenia, reached their target, while the other countries exceeded it. However, most of them were below 2768 p.p. (The highest excess over the target value was observed in Sweden with 11.12 p.p.).
- The indicator X_{14} (the degree to which the target share of energy from renewable sources in gross final energy consumption in transport was achieved) fifteen countries did not reach the target of 10% in 2020 (the largest gap was in Greece—4.66%), and twelve countries exceeded this figure (the most significant exceedance was in Sweden—21.85 p.p.).

The above analysis shows that for each indicator adopted for the study, the values of parameters characterising its distribution indicate smaller or larger disparities in this respect in the EU countries in 2020.

4.2 Results of the Classification of EU Countries Using the TOPSIS Method

The analysis of disproportions based on descriptive parameters concerned individual indicators. Therefore, in order to examine the combined impact of these indicators on green energy inequalities in the EU countries, four synthetic measures have been determined by the TOPSIS method:

- z_i—measure calculated on the basis of all indicators,
- z_{si}—measure calculated on the basis of indicators from the first group,

- z_{di}—measure calculated on the basis of indicators from the second group,
- z_{ni}—measure calculated on the basis of indicators from the third group.

Thus, it was possible to identify green energy disproportions in EU countries from the point of view of a set of specific indicators. The values of these measures for individual EU countries (ordered according to non-increasing values of the measure z_i) together with information on the position of the country in terms of the value of a given synthetic measure and its membership in a specific typological group are given in Table 2.

Table 2 shows that Sweden ranks first for all the indicators examined and is also the leader in terms of indicators characterising the current level of production and consumption of energy from renewable sources and the extent to which the targets specified in the EU Directive have been achieved. Only in terms of the transformation rate Sweden was ranked 6th, but this is a consequence of the high level of indicators in this group that Sweden already achieved in 2016. Estonia and Denmark ranked second and third, respectively, in this ranking. In the case of Estonia, such a high position was determined by indicators related to the rate of transformation—in this respect, Estonia is the leader in the ranking. In contrast, Denmark is not a leader in any of the rankings, and its high position in the combined ranking was determined by its second position among EU countries for group two indicators (the rate of transformation) and fifth place for group one indicators (the current level of renewable energy production and consumption). Unfortunately, Denmark took only 15th place due to the degree of implementation of the assumed values, although it exceeded the target for the indicator X_{13} in 2020 by 1.681 p.p. and fell just 0.3 p.p. short of the target for the indicator X_{14}. Romania was at the end of the ranking, mainly due to a very slow pace of transformation (the last position in this ranking) and exceeding the target value for the indicator X_{13} by only 0.5 p.p. and not achieving the target value for the indicator X_{14} (1.5 p.p. were missing). Hungary and France came second and third, respectively. The position of Hungary is the result of the penultimate position of this country both in the current level of production and consumption of renewable energy, as well as the pace of transformation. In the case of France, such a distant position was mainly the result of failing to meet the targets stipulated in the EU Directive in 2020 (almost 4% of the target for the indicator X_{13} was below the threshold and 0.8% of the target for the indicator X_{14}). Countries that were in the second typological group (positions 4–13), due to the first group of indicators, ranked from 3 (Latvia) to 21 (Netherlands), due to the second group of indicators—from 3 (Greece) to 19 (Croatia), and due to the third group of indicators—from 2 (Finland) to 24 (Latvia). Also, for countries from the third typological group (positions from 14 to 23), one can see a significant variation in the occupied positions depending on the group of indicators: for the first group of indicators, these countries occupied positions from 2 (Austria) to 27 (Poland), for the second group—from 10 (Belgium) to 25 (Portugal), for the third group—from 6 (Italy) to 26 (Poland).

Table 2 Values of synthetic measures with the position and number of the typological group for EU countries ordered by non-increasing values of the measure z_i

Country	Code	z_i	z_i—rank	z_i—group	z_{si}	z_{si}—rank	z_{si}—group	z_{di}	z_{di}—rank	z_{di}—group	z_{ni}	z_{ni}—rank	z_{ni}—group
Sweden	SE	0.696	1	1	0.844	1	1	0.477	6	2	1.000	1	1
Estonia	EE	0.524	2	1	0.423	8	2	0.699	1	1	0.367	4	2
Denmark	DK	0.456	3	1	0.512	5	1	0.604	2	1	0.236	15	3
Finland	FI	0.419	4	2	0.524	4	1	0.378	13	3	0.411	2	2
Netherlands	NL	0.403	5	2	0.188	21	3	0.579	4	1	0.271	9	3
Greece	EL	0.398	6	2	0.294	16	3	0.581	3	1	0.234	17	3
Ireland	IE	0.384	7	2	0.252	19	3	0.553	5	1	0.210	23	3
Croatia	HR	0.371	8	2	0.425	7	2	0.300	19	3	0.397	3	2
Luxembourg	LU	0.367	9	2	0.378	10	2	0.429	9	2	0.283	8	3
Cyprus	CY	0.359	10	2	0.298	13	3	0.455	8	2	0.256	14	3
Germany	DE	0.351	11	2	0.292	17	3	0.472	7	2	0.231	19	3
Latvia	LV	0.347	12	2	0.527	3	1	0.372	14	3	0.202	24	3
Bulgaria	BG	0.345	13	2	0.244	20	3	0.366	15	3	0.360	5	2
Austria	AT	0.332	14	3	0.551	2	1	0.244	23	3	0.269	11	3
Slovakia	SK	0.311	15	3	0.169	23	4	0.379	12	3	0.270	10	3
Portugal	PT	0.308	16	3	0.502	6	1	0.222	25	4	0.269	12	3
Belgium	BE	0.293	17	3	0.166	24	4	0.390	10	2	0.228	20	3
Italy	IT	0.287	18	3	0.322	11	3	0.259	22	3	0.298	6	2
Malta	MT	0.286	19	3	0.280	18	3	0.326	17	3	0.232	18	3
Spain	ES	0.281	20	3	0.303	12	3	0.326	16	3	0.221	22	3

(continued)

Table 2 (continued)

Country	Code	z_i	z_i—rank	z_i—group	z_{si}	z_{si}—rank	z_{si}—group	z_{di}	z_{di}—rank	z_{di}—group	z_{ni}	z_{ni}—rank	z_{ni}—group
Lithuania	LT	0.273	21	3	0.399	9	2	0.236	24	4	0.236	16	3
Poland	PL	0.269	22	3	0.124	27	4	0.387	11	2	0.171	26	3
Slovenia	SI	0.268	23	3	0.296	15	3	0.289	20	3	0.225	21	3
Czechia	CZ	0.267	24	3	0.150	25	4	0.277	21	3	0.296	7	2
France	FR	0.220	25	4	0.185	22	3	0.301	18	3	0.121	27	4
Hungary	HU	0.219	26	4	0.150	26	4	0.204	26	4	0.260	13	3
Romania	RO	0.194	27	4	0.296	14	3	0.146	27	4	0.182	25	3

Such a large diversity of positions occupied by EU countries in individual groups of indicators prompted the authors to conduct a detailed analysis of the disproportions in each group separately, i.e., based on the values of synthetic measures: z_{si}, z_{di} and z_{ni}.

In the case of the first group of indicators, Sweden came first, followed by Austria and Latvia. In the case of Sweden, as many as three indicators (X_4–X_6) were the highest in the surveyed countries and amounted to 60.124%, 31.854% and 66.381%, respectively, and the X_1–X_3 indicators significantly exceeded the level of 60%. For Austria, the X_3 ratio was 78.204% and was the highest in the EU countries, while the X_1 and X_2 ratios exceeded 80%, and the other ratios were above the average. In Latvia, the value of indicator X_1 was at the level of 99.3%, and the other indicators (except indicator X_5) significantly exceeded the EU average. Considering the average level of z_{si} measure at 0.337, it should be noted that in addition to the three countries mentioned above, seven other countries exceeded this level: Finland, Denmark, Portugal, Croatia, Estonia, Lithuania and Luxembourg. On the other hand, the last places were occupied by three Central and Eastern European countries: the Czech Republic, Hungary and Poland. For the Czech Republic and Poland, the values of all indicators from this group were below the EU average, and in the case of Hungary, only indicator X_5. The spatial differentiation of countries according to the value of the measure z_{si} is shown in Fig. 1.

Figure 1 shows that the countries from the first and second typological groups are located primarily in the northern part of the continent. Most of the countries of Western and Southern Europe have been classified in the third typological group. The

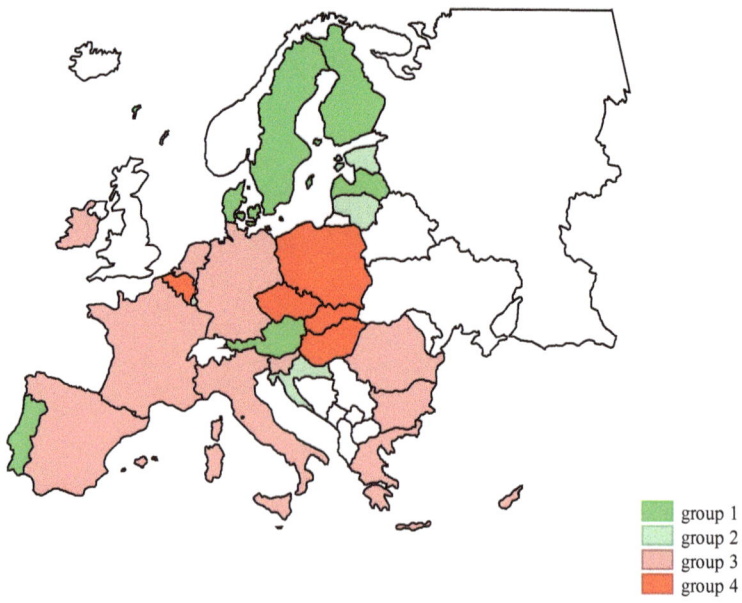

Fig. 1 Spatial differentiation of EU countries by synthetic measure value (z_{si})

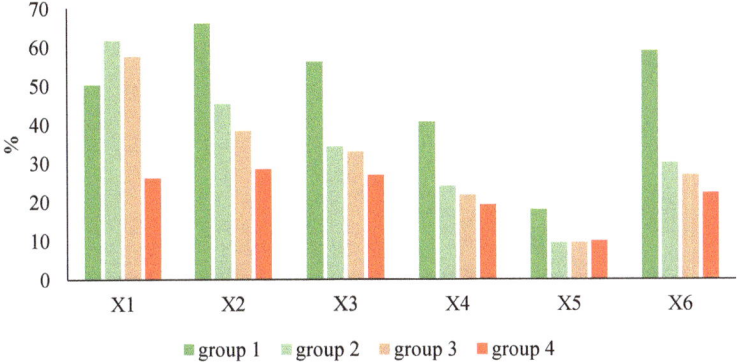

Fig. 2 Average values of indicators X_1–X_6 by typological groups

countries of Central and Eastern Europe are distinct against this background: Poland, the Czech Republic, Slovakia and Hungary are classified into fourth typological groups.

The analysis of the average values of the indicators (Fig. 2), on the basis of which the measure z_{si} was created, shows that the average values for the indicators X_2–X_6 were higher for the countries from the first typological group, which means that in this group of countries (Sweden, Denmark, Finland, Latvia, Austria, Portugal), both electricity production and energy consumption from renewable sources were definitely higher than in other countries. The averages for the indicator X_1 (the share of renewable energy in the total primary energy) are different—the highest average value of this indicator was obtained by the countries from the second typological group, i.e., Estonia, Lithuania, Luxembourg and Croatia. The lowest average values for all diagnostic features were recorded for the last typological group (Poland, the Czech Republic, Slovakia, Hungary, and Belgium).

For the second group of indicators, which characterise the dynamics of the transition to production and consumption of energy from renewable sources by the studied countries in 2020 compared to 2016, the highest value of the measure z_{di} was recorded for Estonia, which in the ranking over-ranked Denmark and Greece. In the case of Estonia, as many as four indicators (X_7–X_9 and X_{11}) were above 10 p.p., including the X_8 indicator, even above 34%. In Denmark, two indicators (X_7–X_8) exceeded 20%, and one indicator (X_9) exceeded 10%, while in Greece, X_7 was higher than 25%, and X_9 exceeded 13%. It demonstrates a very high dynamics in the transformation of energy sources. The lowest value of the measure z_{di} was obtained by: Portugal, Hungary and Romania. In Portugal and Romania, negative values were observed for indicators X_9 and X_{12}, and the remaining indicators only slightly exceeded 2.4 p.p. in Romania and 4.1 p.p. in Portugal. However, in Hungary, negative values were recorded for indicators X_{10} and X_{12}, and the remaining indicators did not exceed 5.7 p.p. The spatial differentiation of countries according to the value of the measure z_{di} is shown in Fig. 3.

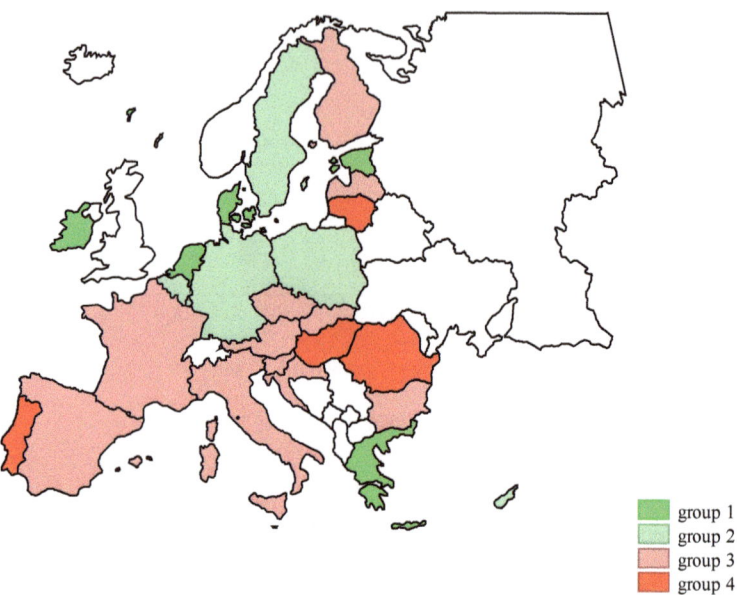

Fig. 3 Spatial differentiation of EU countries by synthetic measure value (z_{di})

Based on Fig. 3, it can be seen that the fastest pace of energy transformation in 2016–2020 was in the northern and central part of the continent, while much slower in the southern and central-eastern European countries. This observation is confirmed by a thorough analysis of the average values of indicators X_7–X_{12} in typological groups (Fig. 4), which shows that the average values of the five indicators X_7–X_9 and X_{11}–X_{12} are higher in the first typological group. Only in the case of the X_{10} indicator (the change in the share of energy from renewable sources in gross final energy consumption in 2020 compared to 2016), a higher average value was obtained for the countries from the second group. At the same time, the negative values of the X_{12} indicator (change in the share of energy from renewable sources in gross final consumption of energy in heating and cooling in 2020 compared to 2016) observed for Romania, Portugal and Hungary, which belong to the fourth typological group together with Lithuania, were confirmed.

The last measure analysed is the measure z_{ni} which allowed for a synthetic evaluation of the extent to which the targets specified for 2020 related to the use of energy from renewable sources, as stipulated in Directive 2009/28/EC, have been achieved. The most successful country in this respect is Sweden, where the target values of indicators X_{13} (the degree of implementation of the target share of energy from renewable sources in the gross final energy consumption) and X_{14} (the degree of implementation of the target share of energy from renewable sources in the gross final energy consumption in transport) were significantly exceeded (the target value for indicator X_{13} was exceeded by more than 11 p.p., for indicator X_{14}—by almost 22 p.p.). The following countries are Finland (X_{13} exceeded by almost 6 p.p., and

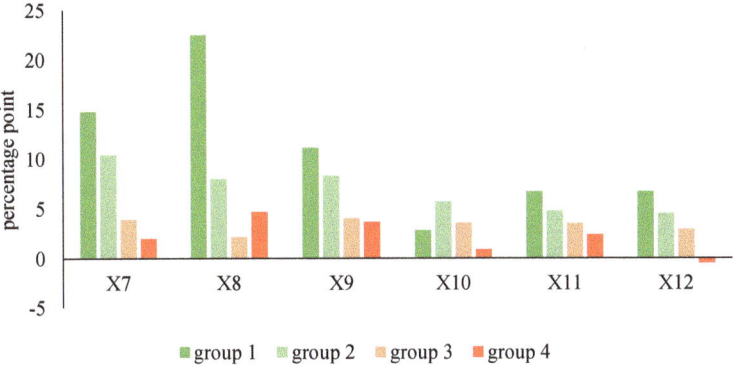

Fig. 4 Average values of indicators X_7–X_{12} by typological groups

X_{14}—by almost 3.5 p.p.) and Croatia (X_{13} exceeded by more than 11 p.p., and in the case of X_{14}, the target was not achieved—3.4 p.p. were missing). The ranking is closed by: Romania, Poland and France. France (the last in the ranking) did not reach any of the targets in 2020, Poland and Romania did not reach the target for the $X14$ indicator, while they slightly exceeded the target for the $X13$ indicator (Poland exceeded this target by 1.1 p.p., and Romania by 0.5 p.p.). The spatial differentiation of EU countries according to the value of the measure z_{ni} is shown in Fig. 5.

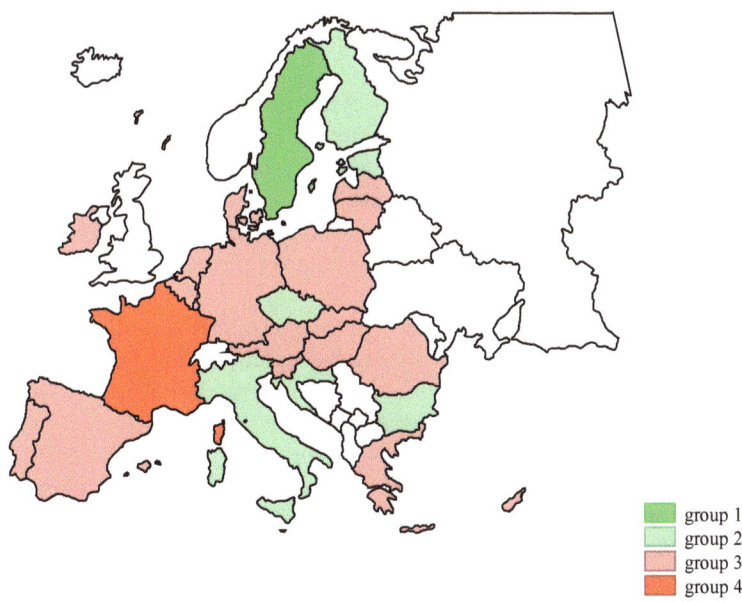

Fig. 5 Spatial differentiation of EU countries by synthetic measure value (z_{ni})

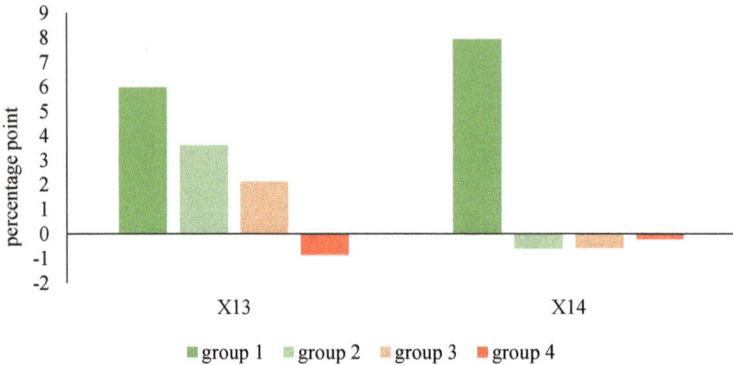

Fig. 6 Average values of indicators X_{13}–X_{14} by typological groups

Figure 5 clearly shows that Sweden and France, due to the degree of achievement of the objectives set out in the EU Directive, form one-element typological groups—the first and the fourth one, respectively. The countries included in the second group are mainly located in the north and south of Europe, as well as in the Balkan Peninsula. The largest group is group 3, which includes countries from different parts of Europe. The second group is dominated by countries that exceeded the target value for the X_{13} indicator by a value ranging from 3.5 p.p. to 7.3 p.p. In comparison, the third group is dominated by countries that exceeded this value by only 3 p.p. or less. In addition, in the third group, most countries did not reach the target for the X_{14} indicator. The average values for both indicators according to typological groups are shown in Fig. 6.

Providing a clear answer to the question of the reasons for the large diversification of EU countries regarding the level of production and consumption of energy from renewable sources is a complex and multifaceted process.

Firstly, the level of production and consumption of renewable energy in the economy depends on technical and environmental conditions. These conditions also determine the diversification of the structure of renewable energy sources according to carriers in individual countries.

Differences in the structure of renewable energy sources according to carriers in EU countries in 2020 were identified using the structures similarity coefficient, in which structures in individual EU countries were compared with the structure for the EU-27, i.e., with the average structure (Fig. 7). Figure 7 shows that biofuels (solid and liquid) were the dominant renewable energy source for the EU-27 in 2020. Nearly 15% of renewable energy was obtained from wind farms, and almost 13% came from hydroelectric power plants. Solar energy provided 7% of renewable energy, and heat pumps and biogas provided 6.3 and 5.6%. However, the least green energy came from geothermal sources (2.9%) and renewable municipal waste (3.9%).

The values of the structural similarity coefficient calculated for individual EU countries in a non-declining order are shown in Fig. 8.

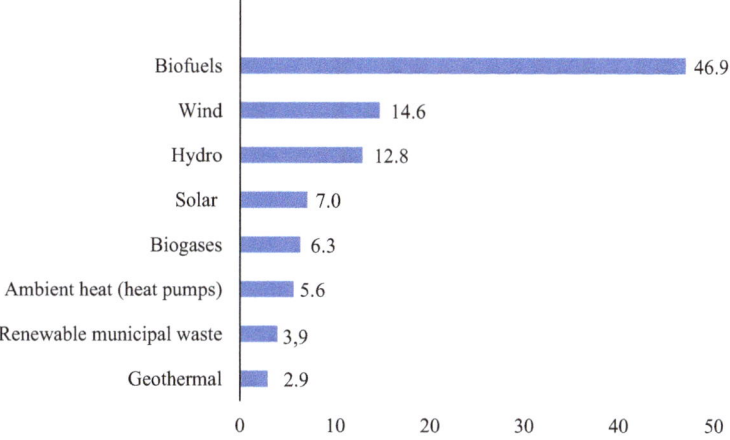

Fig. 7 Structure of renewable energy sources by carriers in the EU-27 in 2020 (%)

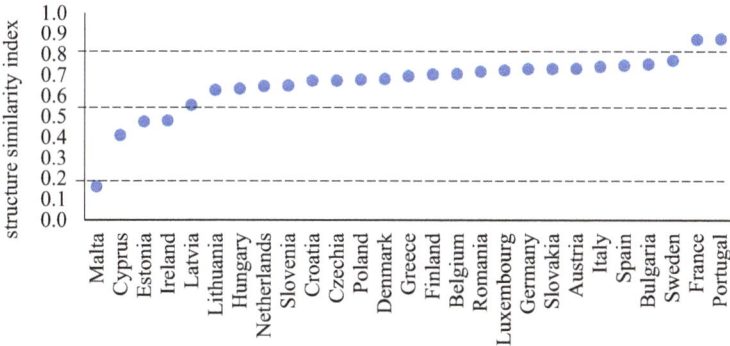

Fig. 8 Coefficient of similarity of renewable energy structures by means in EU countries in relation to the average structure (EU-27) in 2020

On their basis, four groups of countries were distinguished (horizontal lines in Fig. 8), differing in the degree of similarity of the examined structure with the average structure:

- Group 1: France, Portugal;
- Group 2: Belgium, Bulgaria, Czechia, Denmark, Germany, Greece, Spain, Croatia, Italy, Luxembourg, Austria, Poland, Romania, Slovakia, Finland, Sweden, Lithuania, Hungary, Netherlands, Slovenia;
- Group 3: Estonia, Ireland, Cyprus, Latvia;
- Group 4: Malta.

According to the interpretation of the similarity coefficient of structures, in group 1 (the value of the measure of similarity is close to 1), there were countries with a

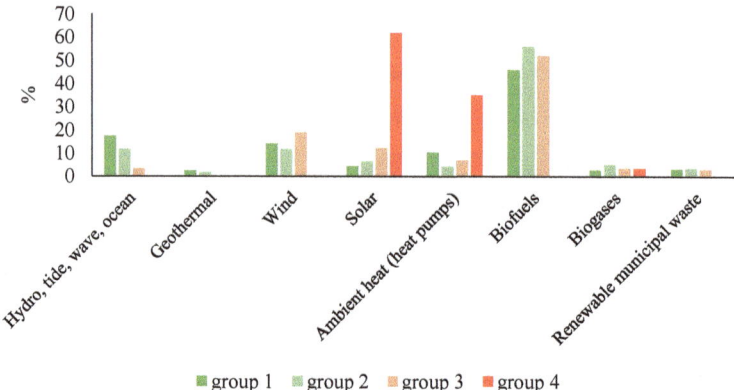

Fig. 9 Structure of the acquisition of primary renewable energy by means in the typological groups of the structures similarity index

structure very similar to the average structure. In contrast, in group 4, Malta has a very different structure compared to the average structure.

For separate groups of countries, the structure of primary renewable energy production (Fig. 9) according to carriers in 2020 was analysed, and thus the specificity of countries in this respect was learned.

It turned out that the dominant renewable energy carrier for groups 1–3 is biofuels, which generated 45.5% (group 1) to 56.0% (group 2) of power. The highest shares for Group 1 countries were recorded for four energy sources: hydroelectric power plants, wind power plants, heat pumps and biofuels. In Group 2 countries, biofuels, water and wind generate the most energy. In the structure of green energy production for countries in group 3, such carriers as wind, solar, and biofuels are particularly prominent. However, in group 4, i.e., Malta, the highest shares were recorded for sun and heat pumps. At the same time, the shares of biogas, geothermal energy and energy from municipal waste in all separate groups were marginal.

Secondly, the above aspect is very important, especially in securing an adequate energy supply level. It is due to the biggest drawback of renewable energy, i.e., unpredictability as to the maximum available power. It depends mainly on the current climatic conditions, e.g., sunlight, temperature and wind force. Therefore, nowadays, "green energy" must now be complemented by alternative carbon or low-carbon energy sources in most countries. Figure 10 presents the structure of primary energy sources (emission and non-emission) divided into typological groups determined based on a synthetic measure of the level (z_{si}).

The analysis of the above graph shows that in the case of group 1 countries, despite the high share of zero-emission energy sources in the total primary energy generated, it has to be supplemented with energy from low-emission sources (nuclear energy) and to a low extent from emission sources: oil and its products and natural gas. For group 2, the share of "green energy" is also high, but it is mainly supplemented by emission sources: bituminous shale, natural gas and oil. Group 3 countries obtain

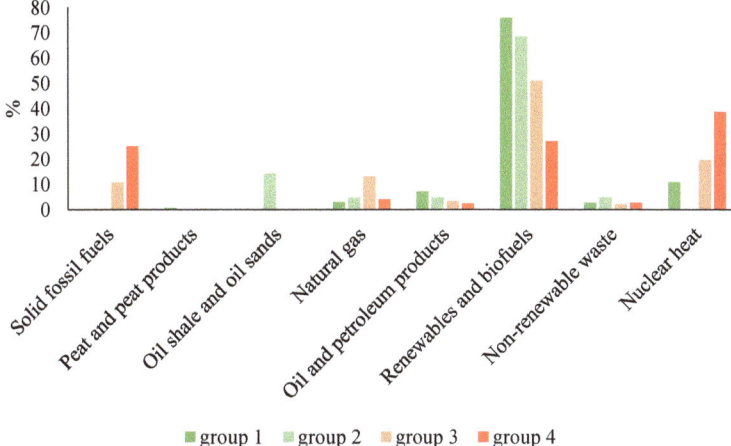

Fig. 10 Structure of primary energy sources according to typological groups determined on the basis of the level measure (z_{si})

half of their energy from renewable sources, and the other half is generated from both low-carbon and emission sources. For Group 4 countries, "green energy" accounts for less than 50% of the total primary energy produced. A high share of nuclear energy and fossil fuels was recorded for this group of countries.

Thirdly, the signing of the Paris Agreement, as well as the targets set out in Directive 2009/28/EC of the European Parliament and the Council, forced rapid changes in the energy carrier of individual EU Member States. The analysis of the rate of changes in the use of primary energy sources in the years 2016–2020 conducted for typological groups distinguished based on a synthetic measure for dynamics (z_{di}) (Fig. 11) shows that in the case of group 1 countries, a clear and rapid departure from emission energy sources, including primarily natural gas and their transition to emission-free substitutes is evident.

Group 2 countries have only slightly reduced the energy supply level from fossil sources and nuclear energy. At the same time, in the analysed period, there was a noticeable increase in the share of energy from renewable sources, while it was higher than the recorded reductions from other sources. It suggests that these sources were additional available capacity in their economies. In the case of the third group of countries, there was a slight reduction in the use of emission energy sources, mainly fossil fuels and natural gas, while increasing the share of renewable energy sources and nuclear energy. The last typological group included countries where there was a slight increase in the share of energy from "green sources" and nuclear energy, with a slight reduction in the share of fossil fuels in the energy mix.

Fourthly, achieving the target shares of energy from renewable sources in gross final energy consumption and transport required action from the Member States. Due to differences in their energy mixes, these activities had to be selected on a case-by-case basis. The analysis of changes in the structure of the use of primary energy

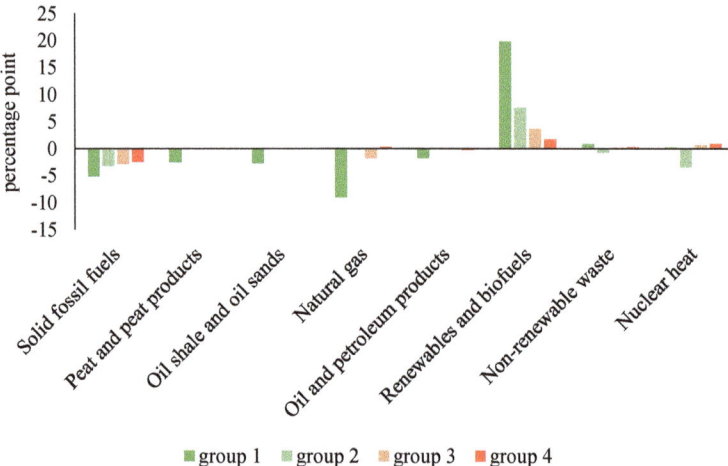

Fig. 11 Change in the structure of primary energy sources in 2016–2020 according to typological groups determined on the basis of the measure for dynamics (z_{di})

sources in the years 2016–2020 carried out for typological groups distinguished on the basis of the measure for standards (z_{ni}) (Fig. 12) indicates that for Sweden forming the first typological group, these changes were subject to a shift away from low-emission nuclear energy and a transition to emission-free energy sources.

On the other hand, for groups 2 and 3, the approach to the assumed levels took place through a clear reduction in the consumption of fossil fuels. At the same time,

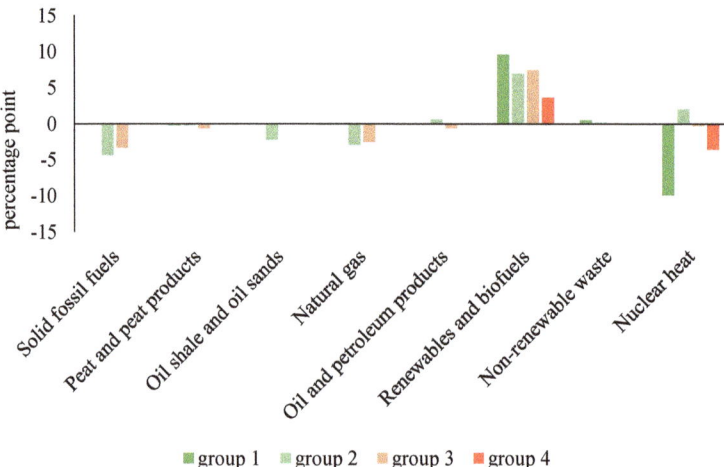

Fig. 12 Change in the structure of primary energy sources between 2016 and 2020 by typological groups determined by the measure for standards (z_{ni})

there was an increase in the use of nuclear energy for the second group. The slight increase in the share of primary energy from renewable sources in the case of France, which forms typological group 4, together with the reduction in the use of nuclear energy, indicates that this country is in the process of slowly abandoning this type of energy.

5 Summary

The research shows that the countries of the European Union are very diverse in terms of production and consumption of energy from renewable sources. In view of the three areas analysed in this study, the three countries of Northern Europe (Sweden, Denmark, Finland) and the two Baltic countries (Latvia, Lithuania) were clearly at the forefront in this respect in 2020. It is due to both cultural conditions (Mikalauskas and Mikalauskienė 2017; Sokołowski 2020; Sovacool and Griffiths 2020) and a responsible approach to the issue of sustainable development, in which green energy is a vital aspect (ESS 2018; Cling et al. 2019). Sweden (the EU's green energy leader) is worth noting, as it is slowly abandoning nuclear power despite its low-carbon nature (World-Nuclear 2022). The worst situation in the field of green energy is observed in Central and Eastern Europe countries, which are still heavily dependent on fossil energy sources, including, above all, coal.

Based on the analysis of the current level of production and consumption of energy from renewable sources and the analysis of changes in this level in 2020 compared to 2016, a group of countries was identified, including France and Spain, which use nuclear energy. Thus, the development of renewable energy sources is much slower. It could be due to the high expenditures incurred to build the power plant (Boccard 2014; WNA 2021a) and environmental problems related to their closure (WNA 2021b; OECD 2016).

The countries of Central and Eastern Europe, despite the lower level of production and consumption of energy from renewable sources, were characterised by a relatively fast pace of its development in the analysed period. It can be caused not only by obligations stemming from international agreements and treaties but also by the desire to diversify energy sources, which results in changes in their energy balance. The effect of accelerating the rate of development of green energy in these countries is a noticeable reduction in the share of emission sources of energy in the production of primary electricity and their replacement with low-carbon and renewable sources. In the context of current events in Europe, such a path seems appropriate, as it protects the economies of these countries from the negative consequences of unpredictable events.

The study results also show that it is no longer possible to abandon the development of renewable energy sources. As a result, there are several reasons for this, the most important of which are the rising costs of fossil fuels (Climat Action 2021;

Jaller-Makarewicz 2021) and the constantly increasing demand for energy (Wolfram et al. 2012; IEA 2021). The latter is particularly relevant for developing economies (Akanonu 2019).

References

Akanonu P (2019) Energy consumption in the developing world by 2040: implications and goals. https://impakter.com/energy-consumption-in-the-developing-world-by-2040-implications-and-goals/. Accessed 28 April 2022

AmCham (2022) Energy transformation in Poland. American Chamber of Commerce in Poland. https://amcham.pl/news/energy-transformation-poland. Accessed 28 April 2022

Asongu S, Odhiambo N (2020) Inequality, finance and renewable energy consumption in sub-Saharan Africa. MPRA 107510. https://mpra.ub.uni-muenchen.de/107510/. Accessed 28 April 2022

Bąk I, Wawrzyniak K, Oesterreich M (2022) Competitiveness of the regions of the European Union in a sustainable knowledge-based economy. Sustainability 14(7):3788. https://doi.org/10.3390/su14073788

Belev G (2021) Regional disparities and features of solar and wind energy. Potential of Bulgaria. Int J Oper Manag 2(1):7–11. https://doi.org/10.18775/ijom.2757-0509.2020.21.4001

Boccard N (2014) The cost of nuclear electricity: France after Fukushima. Energy Policy 66:450–461. https://doi.org/10.1016/j.enpol.2013.11.037

CEF (2022) What is renewable energy? https://www.conserve-energy-future.com/advantages-and-disadvantages-of-renewable-energy.php. Accessed 28 April 2022

Changyong L, Yuanfei L, Qiyuan C, Pengjia S, Malin S, Wei W (2021) Evaluation on the cost of energy transition: a case study of Fujian, China. Front Energy Res 9. https://doi.org/10.3389/fenrg.2021.630847

Chomątowski S, Sokołowski A (1978) Taksonomia struktur. Przegląd Statystyczny 2:217–226

Churchill SA, Ivanovski K, Munyanyi ME (2021) Income inequality and renewable energy consumption: time-varying non-parametric evidence. J Clean Prod 296(11):126306. https://doi.org/10.1016/j.jclepro.2021.126306

Climat Action (2021) New report shows soaring fossil fuel costs have driven 86% of UK electricity price increases. https://www.climateaction.org/news/new-report-shows-soaring-fossil-fuel-costs-have-driven-86-of-uk-electricity. Accessed 28 April 2022

Cling JP, Eghbal-Teherani S, Orzoni M, Plateau C (2019) The differences between EU countries for sustainable development indicators: it is (mainly) the economy! INSEE, G2019/06. https://www.insee.fr/en/statistiques/fichier/version-html/4204820/G2019-06.pdf. Accessed 28 April 2022

De Laurentis C, Pearson PJG (2021) Policy-relevant insights for regional renewable energy deployment. Energy, Sustain Soc 11. https://doi.org/10.1186/s13705-021-00295-4

ESS (2018) European attitudes to climate change and energy: topline results from round 8 of the European Social Survey. https://www.europeansocialsurvey.org/docs/findings/ESS8_toplines_issue_9_climatechange.pdf. Accessed 28 April 2022

EU (2009) Directive 2009/28/EC of the European Parliament and of the Council of 23 April 2009 on the promotion of the use of energy from renewable sources and amending and subsequently repealing Directives 2001/77/EC and 2003/30/EC. https://eur-lex.europa.eu/legal-content/EN/TXT/?uri=CELEX:32009L0028. Accessed 28 April 2022

EUROSTAT (2022) Environment and energy. https://ec.europa.eu/eurostat/databrowser/explore/all/envir?lang=en&display=list&sort=category. Accessed 21 April 2022

Henning H-M, Palzer A (2015) What will the energy transformation cost? Pathways for transforming the German energy System by 2050. Fraunhofer Institute for Solar Energy Systems ISE. https://www.ise.fraunhofer.de/content/dam/ise/en/documents/publications/studies/What-will-the-energy-transformation-cost.pdf. Accessed 28 April 2022

Hwang CL, Yoon K (1981) Methods for multiple attribute decision making. Lecture Notes in Economics and Mathematical Systems 186. https://doi.org/10.1007/978-3-642-48318-9_3

IEA (2021) Global electricity demand is growing faster than renewables, driving strong increase in generation from fossil fuel. https://www.iea.org/news/global-electricity-demand-is-growing-faster-than-renewables-driving-strong-increase-in-generation-from-fossil-fuels. Accessed 28 April 2022

IRENA (2020) The renewable energy transition in Africa. https://www.irena.org/-/media/Files/IRENA/Agency/Publication/2021/March/Renewable_Energy_Transition_Africa_2021.pdf. Accessed 28 April 2022

Jaller-Makarewicz AM (2021) As fossil fuel prices skyrocket globally, renewables grow steadily cheaper. Available via IEEFE. https://ieefa.org/as-fossil-fuel-prices-skyrocket-globally-renewables-grow-steadily-cheaper/. Accessed 28 April 2022

Juszczak, A, Maj M (2020) Rozwój i potencjał energetyki odnawialnej w Polsce. Polski Instytut Ekonomiczny, Warszawa

Kozera A, Łuczak A, Wysocki F (2017) The application of classical and positional TOPSIS methods to assessment financial self-sufficiency levels in local government unit. In: Palumbo F, Montanari A, Maurizio, Vichi M (eds) Data science innovative developments in data analysis and clustering. Studies in classification, data analysis, and knowledge organization. Springer, pp 273–284. https://doi.org/10.1007/978-3-319-55723-6_21

Łatuszyńska M (2013) Multiple-criteria decision analysis using TOPSIS method for interval data in research INTO the level of information society development. Folia Oeconomica Stetinensia 13(21). https://doi.org/10.2478/foli-2013-0015

Layton B (2008) A comparison of energy densities of prevalent energy sources in units of joules per cubic meter. Int J Green Energy 5:438–455. https://doi.org/10.1080/15435070802498036

Markowska M, Strahl D, Sobczak E, Hlaváček P (2019) Podobieństwo struktur zatrudnienia w krajach Unii Europejskiej w latach 2008–2017 – ocena dynamiki. Stud Ind Geogr Comm Polish Geogr Soc 33(4):283–293. https://doi.org/10.24917/20801653.334.17

McGee JA, Greiner PT (2019) Renewable energy injustice: the socio-environmental implications of renewable energy consumption. Energy Res Soc Sci 56:101214. https://doi.org/10.1016/j.erss.2019.05.024

Mikalauskas I, Mikalauskienė A (2017) Cultural acceptance differences of renewable energy technologies. Int J Culture Hist 3(4):270–274. http://www.ijch.net/vol3/111-KS0013.pdf. Accessed 28 April 2022

Młodak A (2006) Analiza taksonomiczna w statystyce regionalnej. Difin, Warszawa

OECD (2016) Costs of decommissioning nuclear power plants. https://www.oecd-nea.org/upload/docs/application/pdf/2019-12/7201-costs-decom-npp.pdf. Accessed 28 April 2022

Pavić Z, Novoselac V (2013) Notes on TOPSIS method. Int J Res Eng Sci 1(2):05–12

Piekut M (2021) The consumption of renewable energy sources (RES) by the European Union households between 2004 and 2019. Energies 14:5560. https://doi.org/10.3390/en14175560

Roszkowska E (2020) Similarity of regions in terms of the structure of the elderly population—proposition of a measure. Optimum Econ Stud 2(100). https://doi.org/10.15290/oes.2020.02.100.11

Sarker BR, Islam KMS (1999) Relative performances of similarity and dissimilarity measures. Comput Ind Eng 37(4):769–807. https://doi.org/10.1016/S0360-8352(00)00011-5

Sharvini SR, Noor ZZ, Chong CS, Lindsay C, Stringer LC, Yusuf RO (2018) Energy consumption trends and their linkages with renewable energy policies in East and Southeast Asian countries: challenges and opportunities. Sustain Environ Res 28(6):257–266. https://doi.org/10.1016/j.serj.2018.08.006

Sinha A (2017) Inequality of renewable energy generation across OECD countries: a note. Renew Sustain Energy Rev 79:9–14. https://doi.org/10.1016/j.rser.2017.05.049

Sokołowski MM (2020) Renewable and citizen energy communities in the European Union: how (not) to regulate community energy in national laws and policies. J Energy Nat Resour Law 38(3):289–304. https://doi.org/10.1080/02646811.2020.1759247

Sovacool BK, Griffiths S (2020) The cultural barriers to a low-carbon future: a review of six mobility and energy transitions across 28 countries. Renew Sustain Energy Rev 119:109569. https://doi.org/10.1016/j.rser.2019.109569

Sweco (2019) Urban energy report. The limits to renewable energy. https://www.swecourbaninsight.com/wp-content/uploads/2020/10/report_the-limits-to-renewable-energy_a4.pdf. Accessed 28 April 2022

UNFCCC (2022) The Paris Agreement. What is the Paris Agreement? https://unfccc.int/process-and-meetings/the-paris-agreement/the-paris-agreement. Accessed 28 April 2022

Uzar U (2020) Political economy of renewable energy: does institutional quality make a difference in renewable energy consumption? Renew Energy 155:591–603. https://doi.org/10.1016/j.renene.2020.03.172

Viñuales JE (2021) Geopolitics of the energy transformation. Governing Globalization 2:148–155. https://geopolitique.eu/en/articles/geopolitics-of-the-energy-transformation/. Accessed 28 April 2022

WNA (2021a) Economics of nuclear power. https://world-nuclear.org/information-library/economic-aspects/economics-of-nuclear-power.aspx. Accessed 28 April 2022

WNA (2021b) Decommissioning nuclear facilities. World-nuclear.org. https://world-nuclear.org/information-library/nuclear-fuel-cycle/nuclear-wastes/decommissioning-nuclear-facilities.aspx. Accessed 28 April 2022

WNA (2022) Nuclear power in Sweden. World-nuclear.org. https://world-nuclear.org/information-library/country-profiles/countries-o-s/sweden.aspx. Accessed 28 April 2022

Wolfram C, Shelef O, Gertler P (2012) How will energy demand develop in the developing world? J Econ Perspect 26(1):119–138. https://pubs.aeaweb.org/doi/pdf/10.1257/jep.26.1.119. Accessed 28 April 2022

Yao X, Yasmeen R, Padda IUH, Shah WUH, Kamal MA (2020) Inequalities by energy sources: an assessment of environmental quality. PLoS ONE 15(3):e0230503. https://doi.org/10.1371/journal.pone.0230503

Zulqarnain M, Saeed M, Ahmad N, Dayan F, Ahmad B (2020) Application of TOPSIS method for decision making R. Int J Sci Res Math Stat Sci 7(2):76–81

Conclusion

The aim of this study was to discuss the issues regarding green energy, its creation and development, as well as address the problems linked with it from the viewpoint of the environment, policy, society, finances, and energy transformation. The results of multi-aspect analyses presented in the chapters allowed the authors to formulate the following conclusions:

1. Europe, in line with the global tendency, is undergoing the transition to renewable sources of energy. EU countries are increasing the share of all RSE, whatever their kind, in the general balance of energy. Such actions contribute to saving natural resources, improving the condition of the natural environment through reducing emissions of pollutants and cutting down the amounts of produced waste, as well as increasing energy security of EU member states.
2. Nowadays, the use of green energy in the economy has become one of the leading topics in the political and social debate conducted on international level, and also in specific integratory groups and within countries. In the EU, the issues related to protection of the environment and RSE are being increasingly treated as a priority, while the regulations concerning environmental protection constitute one of the fastest developing sections of European law. Their common aim is to implement the principles of sustainable development in the global approach, and therefore carrying out the transformation not just in terms of energy sources but also in the broad context of the global economy.
3. The introduction into the market of solutions using green energy is also connected with various problems. The conducted literature review indicates that none of the ways of generating energy is entirely environment-friendly, however the researchers agree that the advantages of generating green energy far outweigh the negative consequences, and the latter occur mostly in the process of production, installation and utilisation of power plants. The low emissions of harmful gases into the atmosphere, and not releasing toxic waste into water reservoirs place the renewable sources of energy on the just side of environmental protection.

4. The progress in the transformation of the traditional economy in the 'green direction' is particularly visible in highly developed countries, which in the process of developing green technologies directly benefit from the advantages linked with their implementation.
5. Green transition will be successful only when companies, governments and individual citizens cooperate in order to achieve global decarbonisation (e.g. through investment, taxation, subsidies, and change of behaviour). In the societal dimension, the key is the change of behaviour, starting from individuals, households, to central governments and entire society. A great challenge rests in ensuring such a direction of socio-economic development in which the demands of green transformation can be seen not merely as a threat or enforced adaptation but also as an opportunity for development.
6. Financing green transformation constitutes one of the most important challenges for the whole modern world, and for individual EU countries. It requires significant investment on the part of the EU, as well as of the domestic public sector and the private sector. The lion's share of the renewable energy projects requires obtaining external financing. The source can be banks and the capital market, which in recent years has been very dynamically developing in the segment of green bonds. The risk involved in the realisation of RSE projects is relatively high, and this also impacts on the cost of capital, hence the growing importance of instruments of the RSE policy, such as tax incentives, loans, guaranteed tariffs, and the Renewable Portfolio Standard (RPS). Their use depends on the type of renewable energy relative to the project, its size and the subject which carries it out.
7. It is very difficult, in fact impossible, to construct one or even a few models of green transformation for EU countries, which at present find themselves at different stages of their development and the approach to the use of green energy sources in the economy. There is also a varied level of social acceptance and the consent in society of the studied countries in respect of incurring higher costs of this transformation.
8. The conducted research suggests that EU countries vary greatly regarding the production and the use of energy from renewable sources. The largest share of such energy was noted among the North European and Baltic countries, while its development was the quickest in Central and Eastern Europe. The lowest level of use of green energy was found in the Balkan countries and in France.
9. The intensive development of renewable energy production globally has already become a necessity, however it requires a reduction or a complete removal of numerous barriers faced by renewable power, among them: social, economic, technological and regulatory. In many countries around the world, especially those less developed, these barriers constitute a significant obstacle to developing technologies of renewable energy and implementing its solutions in order to satisfy the demand for energy.

To sum up, it should be emphasised that the theoretical concepts presented in this study and the results of empirical research are an attempt at the holistic presentation of the problems connected with green energy, yet do not address the problem in its entirety. An important direction of future research should be oriented at identifying the opportunities and barriers linked with implementing green energy in individual EU countries, as well as the related cost–benefit analysis.

GPSR Compliance
The European Union's (EU) General Product Safety Regulation (GPSR) is a set of rules that requires consumer products to be safe and our obligations to ensure this.

If you have any concerns about our products, you can contact us on

ProductSafety@springernature.com

In case Publisher is established outside the EU, the EU authorized representative is:

Springer Nature Customer Service Center GmbH
Europaplatz 3
69115 Heidelberg, Germany

www.ingramcontent.com/pod-product-compliance
Ingram Content Group UK Ltd.
Pitfield, Milton Keynes, MK11 3LW, UK
UKHW021250180426
11946UKWH00003B/55